X Serbian-Bulgarian Astronomical Conference (SBAC 2016)

Belgrade, Serbia
30 May - 3 June 2016

Editors:

Milan S. Dimitrijevic **Milcho K. Tsvetkov**

ISBN: 978-1-5108-3915-1

Printed from e-media with permission by:

Curran Associates, Inc.
57 Morehouse Lane
Red Hook, NY 12571

Some format issues inherent in the e-media version may also appear in this print version.

Printed by Curran Associates, Inc. (2017)

For permission requests, please contact Serbian Astronomical Society
at the address below.

Serbian Astronomical Society
Volgina 7
11000 Belgrade
Serbia

Phone: +381-11-3089-072

mdimitrijevic@aob.bg.ac.rs

Additional copies of this publication are available from:

Curran Associates, Inc.
57 Morehouse Lane
Red Hook, NY 12571 USA
Phone: 845-758-0400
Fax: 845-758-2633
Email: curran@proceedings.com
Web: www.proceedings.com

UDC 520/524(048)

X SERBIAN-BULGARIAN ASTRONOMICAL CONFERENCE

May 30 - June 3, 2016, Belgrade, Serbia

BOOK OF ABSTRACTS

Eds. Milan S. Dimitrijević and Milcho K. Tsvetkov

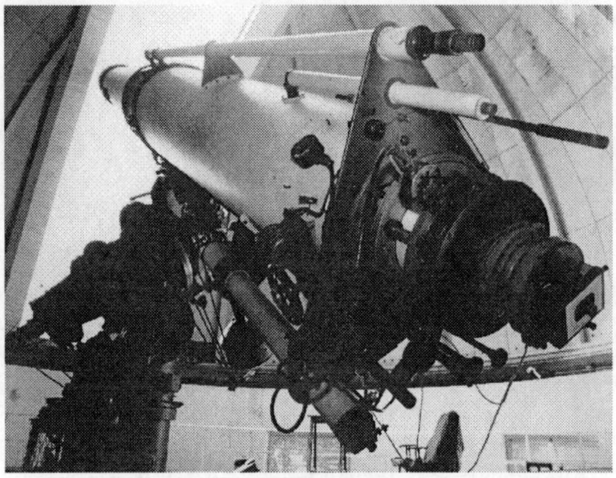

BELGRADE, 2016

ORGANIZERS:
Astronomical Observatory, Belgrade, Serbia
Institute of Astronomy with National Astronomical Observatory, BAS, Bulgaria

Co-organizers:
Faculty of Mathematics and Department of Astronomy, University of Belgrade
Faculty of Civil Engineering, University of Belgrade
Institute of Mathematics and Informatics of the Bulgarian Academy of Sciences

On the front cover: Large Refractor - Zeiss 650/10550 mm of Belgrade Astronomical Observatory.
On the back cover: Moon landscape, Ivanka Ivanova Nikolova.
Text arrangement by computer: Tanja Milovanov

Financially supported by the Ministry of Education, Science and Technological Development of Serbia

Production: Solution+, Bulevar umetnosti 27, 11070 Novi Beograd, in 100 copies

X Serbian-Bulgarian Astronomical Conference (X SBAC)
May 30 - June 3, 2016, Belgrade, Serbia
Book of Abstracts, Eds. M. S. Dimitrijević and M. K. Tsvetkov
Astronomical Observatory, Belgrade, 2016

CONTENTS

INVITED LECTURES

SHORT TALKS

POSTER PAPERS

INVITED LECTURES

X Serbian-Bulgarian Astronomical Conference (X SBAC)
May 30 - June 3, 2016, Belgrade, Serbia
Book of Abstracts, Eds. M. S. Dimitrijević and M. K. Tsvetkov
Astronomical Observatory, Belgrade, 2016

Invited lecture

INVESTIGATION OF SUB-PC AGN POLARIZATION REGIONS AND GRAVITATIONAL LENSES WITH OPTICAL SPECTROPOLARIMETRIC METHOD

Victor L. Afanasiev[1] and Luka Č. Popović[2]

[1]*Special Astrophysical Observatory of the Russian Academy of Sciences, Nizhnij Arkhyz, Karachaevo-Cherkesia 369167, Russia*
[2]*Astronomical Observatory Belgrade, Volgina 7, 11060 Belgrade, Serbia*
E-mail: vafan@sao.ru, lpopovic@aob.rs

Here we present our recent investigation of spectropolarimetric observations of several AGNs with broad emission lines. We found that polarization can be used for investigation of kinematics of the emission gas in the Broad Line Region (BLR) that is connected with the super-massive black hole in the center. This provide us to propose a new method for the super-massive black hole measurements (Afanasiev and Popović, 2015). Also, we present the results obtained from the long-term spectropolarimetric monitoring of two AGNs: Mrk 6 (Afanasiev et al., 2014) and 3C390.3 (Afanasiev et al., 2015). At the end we are going to present preliminarly results obtained from observations and modeling of the spectral polarization in several gravitational lensed objects.

References

Afanasiev, V. L., Popović, L. Č.: 2015, *ApJ*, **800**L, 35.

Afanasiev, V. L., Popović, L. Č., Shapovalova, A. I., Borisov, N. V., Ilić, D.: 2014, *MNRAS*, **440**, 519.

Afanasiev, V. L., Shapovalova, A. I., Popović, L. Č., Borisov, N. V.: 2015, **448**, 2879.

X Serbian-Bulgarian Astronomical Conference (X SBAC)
May 30 - June 3, 2016, Belgrade, Serbia
Book of Abstracts, Eds. M. S. Dimitrijević and M. K. Tsvetkov
Astronomical Observatory, Belgrade, 2016

Invited lecture

CHARACTERISTICS OF THE LONG-TERM SPECTRAL VARIABILITY OF THE AGNs WITH BROAD LINES IN THE OPTICAL SPECTRAL BAND

Nikolaj G. Bochkarev[1], Luka Č. Popović[2] and Alla I. Shapovalova[3]

[1]*Sternberg Astronomical Institute of Lomonosov Moscow State University*
[2]*Astronomical Observatory, Belgrade*
[3]*Special Astrophysical Observatory of the Russian Academy of Science*
E-mail: boch@sai.msu.ru

Active Galactic Nuclei (AGNs) with broad emission lines in the spectra (known as Type 1 AGNs) show variability in all wavelength range (from radio to gamma rays). We use our many-year long spectral and photometric observations of optical variability for investigations of the physics and kinematics of AGN central parts, i.e. the emission regions which are close to the super-massive black hole. Here we give an overview of the results of our analysis of AGN optical spectral variability in a sample of AGNs with broad emission lines.

Invited lecture

SERBIAN – BULGARIAN MINI – NETWORK TELESCOPES: FIRST SIMULTANEOUS OBSERVATIONS OF VARIABLE OBJECTS

S. Boeva[1], G. Damljanović[2], B. Petrov[1], B. Spassov[1], M. Sekulić[2] and G. Latev[1]

[1]*Institute of Astronomy with NAO Rozhen, Bulgarian Academy of Sciences, 72 Tsarigradsko Chausse Blvd., 1784 Sofia, Bulgaria*
[2]*Astronomical Observatory, Volgina 7, 11060 Belgrade, Serbia*
E-mail: sboeva@astro.bas.bg

We report first simultaneous multicolor observations observed in 2015 at 2m and 50/70 Shmidt telescopes of NAO Rozhen, 60 cm telescopes of AO Belogradchik and AS Vidojevica. We present 5-color lightcurves of the cataclysmic variables V425 Cas, V794 Aql, HZ Her and compare the photometric systems of the four telescopes used.

X Serbian-Bulgarian Astronomical Conference (X SBAC)
May 30 - June 3, 2016, Belgrade, Serbia
Book of Abstracts, Eds. M. S. Dimitrijević and M. K. Tsvetkov
Astronomical Observatory, Belgrade, 2016

Invited lecture

R^n GRAVITY AS VIABLE ALTERNATIVE TO DARK MATTER: APPLICATION TO STELLAR DYNAMICS

V. Borka Jovanović[1], P. Jovanović[2], D. Borka[1] and S. Capozziello[3,4,5]

[1]*Atomic Physics Laboratory (040), Vinča Institute of Nuclear Sciences, University of Belgrade, P.O. Box 522, 11001 Belgrade, Serbia*
E-mail: vborka@vinca.rs, dusborka@vinca.rs
[2]*Astronomical Observatory, Volgina 7, P.O. Box 74, 11160 Belgrade, Serbia*
E-mail: pjovanovic@aob.rs
[3]*Dipartimento di Fisica, Universita di Napoli "Federico II", Compl. Univ. di Monte S. Angelo, Edificio G, Via Cinthia, I-80126, Napoli, Italy*
[4]*INFN Sez. di Napoli, Compl. Univ. di Monte S. Angelo, Edificio G, Via Cinthia, I-80126, Napoli, Italy*
[5]*Gran Sasso Science Institute (INFN), Viale F. Crispi, 7, I-67100, L'Aquila, Italy*
E-mail: capozziello@na.infn.it

The role of $f(R)$ gravity, as well as the other modifications of standard Einstein's gravity, is to explain the accelerated expansion, structure formation of the Universe, and some other phenomena at extragalactic scales (such as e.g. flat rotation curves of spiral galaxies) without adding unknown forms of dark energy or dark matter. In $f(R)$ model, the Ricci scalar in the Einstein-Hilbert action is replaced by a general function of it. Its power-law form R^n, here is analyzed using observed orbits of S-stars and also their computer simulations. We review the various consequences of the $f(R)$ gravity parameters (r_c - characteristic radius i.e. scalelength depending on the gravitating system properties and β - universal constant) on stellar dynamics and investigate their constraints from the observed S-star orbits. The presented results show that these observations could put reliable constraints on the parameters of R^n gravity.

X Serbian-Bulgarian Astronomical Conference (X SBAC)
May 30 - June 3, 2016, Belgrade, Serbia
Book of Abstracts, Eds. M. S. Dimitrijević and M. K. Tsvetkov
Astronomical Observatory, Belgrade, 2016

Invited lecture

SERBIAN-BULGARIAN MINI-NETWORK TELESCOPES AND GAIA-FUN-TO

Goran Damljanović[1], Oliver Vince[1], Miljana D. Jovanović[1], Zorica Cvetković[1], Rade Pavlović[1], Svetlana Boeva[2], Georgi Latev[2], Milan Stojanović[1] and Gabrijela Marković[3]

[1]*Astronomical Observatory, Volgina 7, 11060 Belgrade, Serbia*
[2]*Institute of Astronomy with NAO, Bulgarian Academy of Sciences, Tsarigradsko Shosse Blvd. 72, 1784 Sofia, Bulgaria*
[3]*Department of Astronomy, Faculty of Mathematics, University of Belgrade, Studentski trg 16, 11000 Belgrade, Serbia*
E-mail: gdamljanovic@aob.rs

The satellite of Gaia mission was launched at the end of 2013. It is the cornerstone of the European Space Agency (ESA). The Gaia started the astronomical observations in August 2014, and the Gaia Photometric Science Alerts published alerts since October 2014. It is going to map the entire Galaxy, about one billion stars, and near 600 000 quasars (QSOs). About 300 transients have been found: supernovae, cataclysmic variables, candidate microlensing events, etc. Using the 60 cm telescope at Astronomical Station Vidojevica - ASV (of Astronomical Observatory in Belgrade – AOB, Serbia) we observed few objects of Gaia-FUN-TO for the test phase in 2013 and 2014. And at the end of 2014 we continued the observations of Gaia Alerts objects using the Serbian-Bulgarian mini-network telescopes: the 60 cm at ASV, 60 cm at Belogradchik AO, 2 m, 60 cm and 50/70cm Schmidt-camera at Rozhen Observatory (Bulgaria). As result, about 20 objects were observed until the end of 2015 (near 600 CCD images in BVRI filters). Some objects are rare ones as it is the eclipsing AM CVn Gaia14aae one. We observed that object in October 2014 with two telescopes: the 60 cm at ASV and 60 cm at Belogradchik. The paper about that object was published (Campbell *et al.* 2015). Here, we present some of our Gaia-FUN-TO observations and results.

References

Campbell, H. C. et al.: 2015, *MNRAS,* **452**, 1960.

X Serbian-Bulgarian Astronomical Conference (X SBAC)
May 30 - June 3, 2016, Belgrade, Serbia
Book of Abstracts, Eds. M. S. Dimitrijević and M. K. Tsvetkov
Astronomical Observatory, Belgrade, 2016

Invited lecture

A SOLAR PROMINENCE ERUPTION FROM THREE SPACECRAFT POINTS OF VIEW

Momchil Dechev, Ts. Tsvetkov, Kostadinka Koleva and Peter Duchlev

Institute of Astronomy, 72, Tsarigradsko Chaussee Blvd., Sofia, Bulgariass
E-mail: mdechev@astro.bas.bg

We investigated a prominence eruption, occurred between 15:30 and 21:30 UT on May 4, 2014. The eruption was observed by the Solar Dynamic Observatory (SDO) and twin Solar Terrestrial Relations Observatories (STEREO) A and B. The angle between STEREO A and B during the observation provides an unique opportunity to examine the kinematics, structural evolution, and helical twist changes during the eruption. Using the data from the C2 and C3 coronagraphs of Large Angle and Spectrometric Coronagraph (LASCO), an association between the prominence eruption and the spectacular coronal mass ejection was established.

X Serbian-Bulgarian Astronomical Conference (X SBAC)
May 30 - June 3, 2016, Belgrade, Serbia
Book of Abstracts, Eds. M. S. Dimitrijević and M. K. Tsvetkov
Astronomical Observatory, Belgrade, 2016

Invited lecture

PHOTOMETRIC VARIABILITY OF LUMINOUS BLUE VARIABLES IN M33 ON SHORT TIMESCALES

Gantcho Gantchev[1], Petko Nedialkov[1], Valentin Ivanov[2], Evgeni Ovcharov[1], Antoniya Valcheva[1] and Milen Minev[1]

[1]*Department of Astronomy, University of Sofia, 5 J. Bourchier blvd, Sofia 1164, Bulgaria*
[2]*European Southern Observatory, Karl-Schwarzschild-Str. 2, Garching bei München 85748, Germany*
E-mail: japet@phys.uni-sofia.bg

We used SDSS *r*-band aperture photometry and astrometry of ~500 000 stellar-like objects in the M33 galaxy performed by the CASU (Cambridge Astronomy Survey Unit) Astronomical Data Centre in the Institute of Astronomy, University of Cambridge. The observations were carried out with the 2.6m VISTA telescope at the Cerro Paranal, Chile. More than 500 images in that passband were obtained with the OmegaCAM, a large format (16k x16k pixels) CCD camera, and each of them covers a field of view of 1°x1°. The current time span of the data is 2.1 yrs until the end of 2014.

The structure function analysis (Hughes et al. 1992) was applied in order to study the variability of ~30 known or suspected LBVs in the M33 galaxy (Massey et al. 2007) on different time scales. In some cases like Var C the time resolution of the data allows us to confirm an enhanced weekly variations Dm ~ 0.3^{m} which is somehow shorter than the previously know typical monthly variations with the same maximum amplitude thought to be caused by non-radial pulsations.

References

Hughes, P. A., Aller, M. F., Aller, H. D.: 1992, *Astrophysical Journal,* **396**, 469.
Massey, P., McNeill, R. T., Olsen, K. A. G., Hodge, P. W., Blaha, C., Jacoby, G. H., Smith, R. C., Strong, S. B.: 2007, *Astrophysical Journal,* **134**, 2474.

X Serbian-Bulgarian Astronomical Conference (X SBAC)
May 30 - June 3, 2016, Belgrade, Serbia
Book of Abstracts, Eds. M. S. Dimitrijević and M. K. Tsvetkov
Astronomical Observatory, Belgrade, 2016

Invited lecture

ON LONG-TERM CYG X-1=V1357 CYG VARIABILITY

E. A. Karitskaya[1], N. G. Bochkarev[2], V. P. Goranskij[2] and N. V. Metlova[2]

[1]*Institute of astronomy of Russian Academy of Science*
[2]*Sternberg Astronomical Institute of Lomonosov Moscow State University*
E-mail: karitsk@sai.msu.ru

We continue our study of spectral and photometric variability of Cyg X-1 on the base of the 12-year-long series of precision spectral observations and 40-year long series of multicolor photometric observations we have accumulated up to now. In 2006, we revealed for the first time that the temperature of the optical component (O9.7Iab supergiant) was decreasing and its size was increasing over 7 years (1997–2003). Photometry performed at the Crimean Station of the Sternberg Astronomical Institute (Lomonosov Moscow State University) shows that the supergiant's variability on the time scale of decades continues up to now.

X Serbian-Bulgarian Astronomical Conference (X SBAC)
May 30 - June 3, 2016, Belgrade, Serbia
Book of Abstracts, Eds. M. S. Dimitrijević and M. K. Tsvetkov
Astronomical Observatory, Belgrade, 2016

Invited lecture

SPECTRAL CHARACTERISTICS OF THE AGNs TYPE 1 IN THE UV/OPTICAL SPECTRAL BAND

Jelena Kovačević-Dojčinović[1] and Luka Č. Popović[1, 2]

[1]*Astronomical Observatory, Volgina 7, 11060, Belgrade, Serbia*
[2]*Department of Astronomy, Faculty of Mathematics, University of Belgrade, Serbia*
E-mail: jkovacevic@aob.rs

Here we give a review of recent investigations of Active Galactic Nucleus (AGN) spectral characteristics in the UV and optical spectral band. Especially we outline some our results about the narrow/broad line and continuum correlation in the UV/optical spectra in context of the Eigenvector 1 correlations (Boroson and Green 1992) and Baldwin effect, considering an influence of star-burst regions to the AGN emission (Popović and Kovačević 2011). Also, we discuss the intriguing UV/optical Fe II emission lines, for which the mechanism of their excitation and the place of their emission region in AGN structure, are still open questions (Kovačević et al. 2010, Kovačević-Dojčinović and Popović 2015). Some similarities and differences between the optical and UV lines and corresponding continuum will be discussed (see Kovačević et al. 2014, Kovačević-Dojčinović and Popović 2015, Jonić et al. 2016) and proposal for the future work will be given.

References

Boroson, T. A., Green, R. F.: 1992, *ApJS*, **80**, 109.
Jonić, S., Kovačević-Dojčinović, Ilić, D., Popović, L. Č.: 2016, *Ap&SS*, accpeted.
Kovačević, J., Popović, L. Č., Dimitrijević, M. S.: 2010, *ApJS*, **189**, 15.
Kovačević, J., Popović, L. Č., Kollatschny, W.: 2014, *AdSpR*, **54**, 1347.
Kovačević-Dojčinović, J., Popović, L. Č.: 2015, *ApJS*, **221**, 35.
Popović, L. Č., Kovačević, J.: 2011, *ApJ*, **738**, 68.

X Serbian-Bulgarian Astronomical Conference (X SBAC)
May 30 - June 3, 2016, Belgrade, Serbia
Book of Abstracts, Eds. M. S. Dimitrijević and M. K. Tsvetkov
Astronomical Observatory, Belgrade, 2016

Invited lecture

METHODS FOR DOCUMENT IMAGE DE-WARPING

Lasko Laskov

Department of Informatics at New Bulgarian University, Sofia, Bulgaria
E-mail: llaskov@nbu.bg

The problem for design of methods and algorithms that can automatically extract data from digitized astrographic plates (Laskov 2013) incorporates a number of techniques that are usually met in the field of document image processing. One of the problems that arise is the task for automatic software rectification of scanned document pages that suffer from curvature distortion due to the physical folding of the paper that has been captured by a digital camera or a scanner.

In this paper we present some ot the state of the art techniques for document image de-warping. Our goal is to examine the existing methods, and to investigate their possible application in the image processing and pattern recognition techniques used in astrographic plates automatic data extraction.

References

Laskov, L., Tsvetkov Milcho: 2013, "Data Extraction form Carte du Ciel Tripple Images", *Serdica J. Computing* **7(4)**, 317-332.

X Serbian-Bulgarian Astronomical Conference (X SBAC)
May 30 - June 3, 2016, Belgrade, Serbia
Book of Abstracts, Eds. M. S. Dimitrijević and M. K. Tsvetkov
Astronomical Observatory, Belgrade, 2016

Invited lecture

NEWS OF PULSAR ASTRONOMY

Igor F. Malov

Pushchino Radioastronomical observatory,
P. N. Lebedev Physical Institute, Russian Academy of Sciences
E-mail: malov@prao.ru

The known catalog of radio pulsars (Manchester et al. 2005) contains now 2536 objects. There are among them isolated sources, neutron stars in binary systems and globular clusters, pulsars emitting X-ray and gamma-ray radiation. Some classes of anomalous pulsars demand the detailed investigations and understanding of their nature.These are Anomalous X-ray pulsars (AXPs), Soft Gamrra-ray Repeaters (SGRs), X-ray Dim Isolated Neutron Stars (XDINSs), Compact Central Objects in SNRs (CCOs), and Rotation Radio Transients (RRATs). Three type of models (magnetar, accretion and drift) were put forward to explain some peculiarities of these objects. Each of them has its advantages and some specific difficulties. They will be briefly discussed in this report. It was shown that normal pulsars can be divided by two main classes: objects with short (P< 0.1 sec) and long (P>0.1 sec) rotation periods. They will be briefly described. The known gamma pulsars (Abdo et al. 2013) possess very high magnetic inductions B_lc near the light cylinder (Malov and Timirkeeva, 2014). These fields are three order stronger than fields in gamma-quiet radio pulsars. This give the possibility to predict discoveries of new pulsed gamma-ray sources in radio pulsars with high values of B_lc. The current picture of ideas in the field of pulsar investigations will be presented.

References

Abdo, A. A.., Ajello, M., Allafort, A. et al.: 2013., *Astron. Astrophys. Suppl. Ser.*, **208**, 17.

Malov, I. F., Yimirkeeva, M. A.: 2015, *Astron. Rep.*, **59**, 865.

Manchester, R. N., Hobbs, G. B., Teoh, A., Hobbs, M.: 2005, *Astron. J.*, **129**, 1993.

X Serbian-Bulgarian Astronomical Conference (X SBAC)
May 30 - June 3, 2016, Belgrade, Serbia
Book of Abstracts, Eds. M. S. Dimitrijević and M. K. Tsvetkov
Astronomical Observatory, Belgrade, 2016

Invited lecture

IONOSPHERIC PERTURBATIONS INDUCED BY SOLAR X-RAY FLARES

Aleksandra Nina

Institute of Physics, University of Belgrade, Belgrade, Serbia
E-mail: sandrast@ipb.ac.rs

The solar X-ray flare as one of the most intensive perturbers of the ionosphere induces variations in parameters of plasma in this part of the atmosphere (Nina *et al.* 2012, Nina and Čadež 2014) and, consequently, affects propagation of telecommunication signals (Bajčetić *et al.* in press). Keeping in mind that propagation of electromagnetic waves in some medium depends primarily on the ambient electron density distribution, the analysis of induced variations of properties of this plasma parameter is very important in investigations of space weather influence on telecommunications.

This work presents a study of variations in electron density characteristics in the ionospheric D-region during a solar X-ray flare based on monitoring the low ionosphere by the 23.4 VLF radio signal emitted in Germany and recorded in Serbia.

References

Bajčetić, J., Nina, A., Čadež, V. M., Todorović, B. M.: *Thermal Science,* in press, doi:10.2298/TSCI141223084B.

Nina, A., Čadež, V. M., Srećković, V., Šulić, D.: 2011, *Baltic Astronomy,* **20**, 609.

Nina, A., Čadež, V. M.: 2014, *Advances in Space Researc,* **54**(7), 1276.

Invited lecture

RELATIONSHIP BETWEEN SOLAR MAGNETIC FIELDS OF DIFFERENT SPATIAL SCALES AND SOME CONTROVERSIAL PROBLEMS OF THE DYNAMO THEORY

V. Obridko and I. V. Dmitrieva

IZMIRAN, Troitsk, Moscow, Russia
E-mail: obridko@mail.ru

SOHO/MDI magnetograms obtained in the epoch of maximum of Cycle 23 are compared with 284 A solar images from the SOHO/EIT space telescope. The analysis has corroborated the existence of complexes of activity that involve active regions (AR) and equatorial coronal holes (CH). Both AR and CH are embedded in an extended magnetic region dominated by the magnetic field of one sign, but not strictly unipolar. Moreover, this magnetic region can only be isolated by smoothing over about 80 arc sec. It is suggested that the solar magnetic field consists of three components that are interrelated but are generated by different mechanisms. The main (or background) magnetic field has a strength of about 3 G, and it can be detected by reasonable smoothing and seems to provide the basis for the occurrence of coronal holes and open magnetic fields. The background field is superimposed by a small-scale magnetic field with the mean intensity of 18 G, which strongly fluctuates both in magnitude and in sign. The total flux of the fields lower than 100 Gs does not have cyclic variation. The occurrence of the latter requires an additional mechanism, which can be described as subsurface small-scale dynamo. And finally, there is a third mechanism, which enhances the field and controls the fields of active regions (i.e., the fields of sunspots and faculae). Although the fields here are much stronger (from 40 G up to 3000 G), their contribution to the total flux is not large 30-40%). These fields are extremely variable; they can change by 10-30% for one or two days. Since the flux from these fields is much smaller than that from the main field, the scenario of regeneration of the latter proposed by Babcock and Leighton seems doubtful.

X Serbian-Bulgarian Astronomical Conference (X SBAC)
May 30 - June 3, 2016, Belgrade, Serbia
Book of Abstracts, Eds. M. S. Dimitrijević and M. K. Tsvetkov
Astronomical Observatory, Belgrade, 2016

Invited lecture

SCANNING MOSCOW PLATE COLLECTION: PROGRAM, ELECTRONIC CATALOGS, AND NEW VARIABLE STARS IN THE FIELD OF 104 HERITLE

N. N. Samus[1,2], S. V. Antipin[2,1], D. M. Kolesnikova[1], K. V. Sokolovsky[3,2] and M. K. Tsvetkov[4]

[1]*Institute of Astronomy, Russian Academy of Sciences*
[2]*Sternberg Astronomical Institute, Lomonosov Moscow University*
[3]*IAASARS, National Observatory of Athens, Greece*
[4]*Institute of Mathematics and Informatics, Bulgarian Academy of Sciences*
E-mail: samus@sai.msu.ru

Astronomical plate archives contain a wealth of information about changes in positions and brightness of celestial objects on timescales of decades and areuseful as the first epoch for comparison with modern-day observations. We describe the astronomical plate collection of the Sternberg AstronomicalInstitute (Lomonosov Moscow University) and the program of its digitization using high-resolution scanners. The program is focused on the search for new variable stars. After scanning all plates for a given plate center, we perform source extraction, photometric/astrometric calibration and construct light curves of all objects in the field using a custom-made pipeline based on open-source software. Variable stars identification is performed by running a period search on the constructed light curves followed by human-eye inspection of identified candidates. Only a small part of the plate collection hasbeen digitized so far. We present the recently obtained results for a field in Hercules with the center at 104 Her (18h11m54.2s +31d24'19", 2000.0). About 300 new variable stars have been discovered, some of thcm showing unusual fcaturcs.Thc most interesting variables we found will be discussed in our talk. Access to plate logs of the scanned parts of our collection is provided; interested users can get access to plate scans they need upon request. We will incorporate the corresponding plate logs into WFPDB.

X Serbian-Bulgarian Astronomical Conference (X SBAC)
May 30 - June 3, 2016, Belgrade, Serbia
Book of Abstracts, Eds. M. S. Dimitrijević and M. K. Tsvetkov
Astronomical Observatory, Belgrade, 2016

Invited lecture

ATOM-RYDBERG ATOM PROCESSES IN THE STELLAR ATMOSPHERES: DWARF ATMOSPHERES, QUIET SUN AND SUNSPOTS

Vladimir A. Srećković[1], Anatolij A. Mihajlov[1], Ljubinko M. Ignjatović[1] and Milan S. Dimitrijević[2,3,4]

[1]*University of Belgrade, Institute of Physics, P. O. Box 57, 11001, Serbia*
[2]*Astronomical Observatory, Volgina 7, 11160 Belgrade 74, Serbia*
[3]*IHIS-Technoexperts, Bežanijska 23, 11080 Zemun, Serbia*
[4]*Observatoire de Paris, 92195 Meudon Cedex, France*
E-mail: vlada@ipb.ac.rs

The rate coefficients of the chemi-ionization processes in H(1s) + H*(n,l) and He(1s2) + He*(n,l) collisions, taking into account the influence of the corresponding (n-n')-mixing processes, are determined for the temperature range characteristic for the solar and DB white-dwarf atmospheres.

The theory and modeling is done on the basis of the method described in Mihajlov *et al.* (2015). The direct calculations of the chemi-ionization rate coefficients are carried out for the models of the quiet Sun, sunspot and DB white-dwarf atmospheres.

It is demonstrated that the inclusion of (n-n') mixing processes in the calculation influences the values of chemi-ionization rate coefficients significantly in all examined cases.

References

Mihajlov, A. A., Srećković, V. A., Ignjatović, Lj. M., Klyucharev, A. N., Dimitrijević, M. S., Sakan, N. M.: 2015, *Journal of Astrophysics and Astronomy,* **36**, 3.

X Serbian-Bulgarian Astronomical Conference (X SBAC)
May 30 - June 3, 2016, Belgrade, Serbia
Book of Abstracts, Eds. M. S. Dimitrijević and M. K. Tsvetkov
Astronomical Observatory, Belgrade, 2016

Invited lecture

ON SOME BAMBERG WIDE-FIELD PLATE CATALOGUES RECENTLY INCORPORATED INTO WFPDB

K. Tsvetkova[1], M. Tsvetkov[1], N. Kirov[2], D. Kalaglarsky[1], H. Edelmann[3] and U. Heber[3]

[1]*Institute of Mathematics and Informatics, Bulgarian Academy of Sciences, Acad. Georgi Bonchev Str., Block 8, 1113 Sofia, Bulgari*
[2]*New Bulgarian University, Montevideo 21, 1618 Sofia, Bulgaria*
[3]*Dr. Karl Remeis-Observatory, Astronomical Institute of the Erlangen-Nuremberg University, Sternwartstrasse 7, D-96049 Bamberg, Germany*
E-mail: katya@skyarchive.org

We present the incorporation of some Bamberg wide-field plate catalogues in the Wide-Field Plate Database (WFPDB), recently prepared in the framework of a project for astronomical photographic plate preservation funded by the German Science Foundation. These catalogues include 8400 plates obtained with two Tessar cameras (with WFPDB identifier BAM003 and BAM009A), a Xenon camera (BAM006), a Vierlinser camera (BAM008), an Ernostar camera (BAM009B), and a Dogmar camera (BAM011) in the period 1931 – 1963. The plate observations aimed at investigations of variable stars in the Northern sky. Some of the cameras were first mounted on an astrograph in Bamberg in preparation for and for testing of the Bamberg large-scale project for variable star research in the Southern sky in the period 1963 – 1976, mostly from Boyden and Mount John observatories.

The WFPDB data format puts requirements on content and data structure in the WFPDB. Therefore the original data of these Bamberg plate catalogues were converted to the needed data format by the WFPDB software tools for time and coordinates conversions.

Currently information on about 30 Bamberg plate archives done in the frameworks of the Bamberg Northern Sky Survey and Bamberg Southern Sky Survey can be found in the WFPDB, containing 34500 wide-field photographic astronomical plates.

Invited lecture

NEW DETERMINATION OF PERIOD AND QUALITY FACTOR OF CHANDLER WOBBLE, CONSIDERING GEOPHYSICAL EXCITATIONS

Jan Vondrák[1], Cyril Ron[1] and Yavor Chapanov[2]

[1]*Astronomical Institute CAS, Boční II, 14100 Prague, Czech Republic*
[2]*National Inst. Geophys., Geod. and Geogr. BAS, Acad. G. Bonchev 3, 1113 Sofia, Bulgaria*
E-mail vondrak@ig.cas.cz

Polar motion consists of both free (Chandler Wobble) and forced components. The latter are caused by different excitations of geophysical origin. Very long-periodic (or secular) part is most probably due to post-glacial rebound, shorter periodic part (with dominant annual period) are caused mainly by motions of the atmosphere and oceans. Recently it was also proposed that impulse-like excitations due to geomagnetic jerks might be responsible for rapid changes of the amplitude and phase of Chandler wobble. In order to precisely determine the parameters of the free part, it is necessary to consider all these influences. The result is the new determination of the period and quality factor of Chandler wobble, free from additional geophysical excitations.

Invited lecture

GAIA SCIENCE ALERTS FOLLOW-UP OBSERVATIONS

Werner W. Zeilinger

Department of Astrophysics at the University of Vienna, Tuerkenschanzstrasse 17, 1180 Wien, Austria
E-mail: werner.zeilinger@univie.ac.at

Results of follow-up observations of Gaia Science Alerts obtained with the University of Vienna 1.5m telescope are presented.

SHORT TALKS

X Serbian-Bulgarian Astronomical Conference (X SBAC)
May 30 - June 3, 2016, Belgrade, Serbia
Book of Abstracts, Eds. M. S. Dimitrijević and M. K. Tsvetkov
Astronomical Observatory, Belgrade, 2016

Short talk

ASTRONOMICAL SOCIETY "RUDJER BOSKOVIC"

Jovan Aleksić

University of Belgrade, Faculty of Mathematics, Department of Astronomy
E-mail: jaleksic@aob.rs

Astronomical Society "Rudjer Boskovic" is the oldest one in the Balkans. It was founded in 1934, and its main goal is popularization of astronomy.

The Society is located in two buildings, Public Observatory and Planetarium. Both of them, as well as the society itself, celebrated anniversaries in 2014 - the Society 80 years, Public Observatory 50 years and Planetarium 45 years since the establishment.

In this paper, a brief history of the Society and its activities are presented.

X Serbian-Bulgarian Astronomical Conference (X SBAC)
May 30 - June 3, 2016, Belgrade, Serbia
Book of Abstracts, Eds. M. S. Dimitrijević and M. K. Tsvetkov
Astronomical Observatory, Belgrade, 2016

Short talk

SERBIAN VIRTUAL OBSERVATORY: BELGRADE ASTROPHOTOGRAPHIC PLATE ARCHIVE OF SOLAR OBSERVATIONS

Vladimir Benishek and Vojislava Protitch-Benishek

Astronomical Observatory, 11060 Belgrade, Volgina 7, Serbia
E-mail: vlaben@yahoo.com

Between 1935 and 1997 more than 15000 high quality negatives of various celestial objects were obtained by four telescopes of Belgrade Astronomical Observatory. Among them, we have to emphasize that more than 4000 systematic photographic observations of the solar photosphere were performed. During the AGI and ACI (1957-1959) 839 days of observations, with 2885 plates of solar disc were performed with Zeiss Refractor 65/1050.

As in the case of Belgrade Wide Field Database which is just prepared as the computer – readable Catalog, we started to digitize the images of the Sun. It will be a new scientific Project which will allow efficient computer base analysis of the data set.

References

Đurković, P. M., Protić, M.B., Arsenijević, J., Kubičela, A., Kovačević, O., Grujić, R., Dačić, Lj., Đokić, M., Čepinac, Č.: 1965, OBSERVATIONS DU SOLEIL en 1957, 1958 et 1959, *Publ. L'Obs. Astron. de Beograd,* **11**.

Protitch-Benishek, V., Benishek, V.: 2011, Serbian Virtual Observatory: Current Status of Belgrade Wide Field Plate Archive. Future Plans and Topics, *JENAM 2011,* Book of Abstracts, 173.

X Serbian-Bulgarian Astronomical Conference (X SBAC)
May 30 - June 3, 2016, Belgrade, Serbia
Book of Abstracts, Eds. M. S. Dimitrijević and M. K. Tsvetkov
Astronomical Observatory, Belgrade, 2016

Short talk

LITHIUM, SODIUM, AND POTASSIUM RESONANCE LINES PRESSURE BROADENED BY HELIUM ATOMS

Robert Beuc[1], Gillian Peach[2], Mladen Movre[1] and Berislav Horvatić[1]

[1]*Institute of Physics, HR-10000 Zagreb, Croatia*
[2]*University College London WC1E 6BT, UK*
E-mail: beuc@ifs.hr

The resonance lines of the alkali-metal atoms, broadened by collisions with helium atoms, are prominent features in the spectra of brown dwarfs.

We carried out calculations of the emission and absorption spectra in the far red and blue wings of the first resonant doublets of lithium, sodium and potassium perturbed by helium atoms, for temperatures from 500 to 3000 K.

For our calculation we used carefully constructed electronic adiabatic potential energy curves and electronic transition dipole moments of Li-He, Na-He and K-He molecules. The alkali-He molecule is treated in a three-body model, as a He atom and an alkali ion plus a valence electron, where the electron-atom and electron-atomic ion interactions are represented by model potentials (Peach 1982, Mullamphy *et al.* 2007).

In the computation of the emission and absorption coefficients, we used full quantum mechanical calculation on the Fourier grid, where vibrational wave functions are represented on a finite number of uniformly spaced grid points (Beuc et al 2014, Horvatić et al 2015). Within this scheme, all transitions between bound, quasibound and free vibrational states are included in the same way.

The obtained absorption and emission spectra are compared with previous theoretical calculations.

References

Beuc, R., Movre, M., Horvatić, B.: 2014, *Eur. Phys. J. D*, **68**, 59.
Horvatić, B, Beuc, R., Movre, M.: 2015, *Eur. Phys. J. D*, **69**, 113.
Peach, G.: 1982 *Comments At. Mol. Phys.,* **11,** 101.
Mullamphy, D. F. T., Peach, G., Venturi, V., Whittingham, I. B., Gibson, S. J.: 2007, *J. Phys. B: At. Mol. Opt. Phys.,* **40,** 1141.

X Serbian-Bulgarian Astronomical Conference (X SBAC)
May 30 - June 3, 2016, Belgrade, Serbia
Book of Abstracts, Eds. M. S. Dimitrijević and M. K. Tsvetkov
Astronomical Observatory, Belgrade, 2016

Short talk

THE TWO COMPONENT MODEL OF THE BLR OPTICAL EMISSION IN THE NGC 5548

Edi Bon[1], Paola Marziani[2] and Nataša Bon[1]

[1]*Astronomical Observatory, Volgina 7, Belgrade, Serbia*
[2]*INAF, Padova, Italy*
E-mail: ebon@aob.rs

It is widely accepted that the enormous emission of the active galactic nuclei (AGN) originates in accretion processes around their central supermassive black holes. The NGC 5548 monitoring supports the idea that the BLR is made of two sub-region: one emitting the broader line base (VBLR) that we show to be associated with the accretion disk in NGC 5548; and a second one, emitting the line core (called ILR or BLR by different authors) whose structure remains still unclear at present but that may be associated with a more isotropic distribution of dense gas clouds. We present preliminary results on the accretion disk and ILR/BLR parameters in various activity phases..

X Serbian-Bulgarian Astronomical Conference (X SBAC)
May 30 - June 3, 2016, Belgrade, Serbia
Book of Abstracts, Eds. M. S. Dimitrijević and M. K. Tsvetkov
Astronomical Observatory, Belgrade, 2016

Short talk

A NEW METHOD TO STUDY AGN SPECTRAL VARIABILITY

Nataša Bon[1], Edi Bon[1] and Paola Marziani[2]

[1]*Astronomical Observatory Belgrade, Serbia*
[2]*INAF, Osservatorio Astronomico di Padova, Padova, Italia*
E-mail: nbon@aob.bg.ac.rs

We introduce a new method for analyzing spectral variability of AGN using full fitting of spectra with the ULYSS code. The code has been adopted to fit spectra of Seyfert 1 galaxies with the model that includes a linear combination of non-linear components - AGN continuum, stellar population, broad and narrow emission lines and Fe II templates. We compare results to an analysis made with IRAF's Specfit package, where we used full spectrum model fitting with multiple line components such as stellar templates based on an elliptical galaxy and FeII emission templates based on IZw1. We made consistency tests between the two methods, and we concluded that the new method based on the ULYSS code can be also used in Type I AGN spectral analysis. In this way we are able to obtain a consistent analysis of a type-1 active galaxy during monitoring campaigns. Here we present preliminary results of lags and line shape analysis obtained with our new method.

Short talk

EMISSION PROPERTIES OF WHITE DWARF'S ACCRETING BINARIES BY THE POLARIZATION METHODS

Daniela Boneva

Space Research and Technology Institute, Sofia, Bulgaria
E-mail: danvasan@space.bas.bg

We investigate the properties of emission lines in accreting binaries. Flares activity in two binary stars with accretion disc is considered. We present our modeling on the base of polarization methods. The modeling prediction gives the polarization degree according to their light curves and energy spectrum. Our analysis indicates that the polarization parameters show variations in their values during the flares activity and it mainly depends on radiation properties.

Short talk

CONSTRAINTS ON SANDERS GRAVITY FROM PRECESSION OF ORBITS OF S2-LIKE STARS: A CASE OF A BULK DISTRIBUTION OF MASS

D. Borka[1], P. Jovanović[2], V. Borka Jovanović[1] and S. Capozziello[3,4,5]

[1]*Atomic Physics Laboratory (040), Vinča Institute of Nuclear Sciences, University of Belgrade, P.O. Box 522, 11001 Belgrade, Serbia*
E-mail: dusborka@vinca.rs, vborka@vinca.rs
[2]*Astronomical Observatory, Volgina 7, 11160 Belgrade, Serbia*
E-mail: pjovanovic@aob.rs
[3]*Dipartimento di Fisica, Universita di Napoli "Federico II", Compl. Univ. di Monte S. Angelo, Edicio G, Via Cinthia, I-80126, Napoli, Italy*
[4]*INFN Sez. di Napoli, Compl. Univ. di Monte S. Angelo, Edificio G, Via Cinthia, I-80126, Napoli, Italy*
[5]*Gran Sasso Science Institute (INFN), Viale F. Crispi, 7, I-67100, L'Aquila, Italy*
E-mail: capozziello@na.infn.it

In this paper we investigate possible applications of the observed orbits of S2-like stars around Galactic Center for constraining the gravitational potentials derived from modified gravity models in the absence of dark matter. To this aim, an analytic fourth-order theory of gravity, nonminimally coupled with a massive scalar field, is considered. We study the constraints on the f (R, φ) (Sanders) gravity where R is the Ricci scalar and φ is a scalar field and demonstrate that such constraints could be obtained from the observations of S2-like stars by the present and next generation large telescopes. Our results show that Sanders gravity affects the simulated orbits in the qualitatively opposite way with respect to a bulk distribution of matter (including a stellar cluster and dark matter distributions) in Newton's gravity. This is due to the fact that the extended mass cause the retrograde orbital precession, while Sanders gravity cause the orbital precession in the same direction as predicted by General relativity.

X Serbian-Bulgarian Astronomical Conference (X SBAC)
May 30 - June 3, 2016, Belgrade, Serbia
Book of Abstracts, Eds. M. S. Dimitrijević and M. K. Tsvetkov
Astronomical Observatory, Belgrade, 2016

Short talk

MHD WAVES IN THE LOWER TERRESTRIAL IONOSPHERE

Vladimir M. Čadež[1] and Aleksandra Nina[2]

[1]*Astronomical Observatory of Belgrade, Belgrade, Serbia*
[2]*Institute of Physics, University of Belgrade, Belgrade, Serbia*
E-mail: vcadez@aob.rs

Hydrodynamic waves in the ionosphere influence reflection properties of radio waves emitted and received at the ground of the Earth. The resulting time variations in recorded amplitude of the very low frequency (VLF) radio signal obtained in our earlier investigations indicate the existence of such waves in lover parts of the ionosphere assuming negligible effects of the geomagnetic field (Nina and Čadež 2013).

In this presentation we generalize this issue by inclusion of the geomagnetic field and estimation of its contribution in amplitude time variations of the reflected VLF wave as recorded by the Belgrade station in Institute of Physics in Belgrade, Serbia.

References

Nina A., Čadež V. M.: 2013, *Geophysical Research Letters,* **40** (18), 4803.

X Serbian-Bulgarian Astronomical Conference (X SBAC)
May 30 - June 3, 2016, Belgrade, Serbia
Book of Abstracts, Eds. M. S. Dimitrijević and M. K. Tsvetkov
Astronomical Observatory, Belgrade, 2016

Short talk

EARTH ORIENTATION PARAMETERS AND GRAVITY VARIATIONS DETERMINED FROM LAGEOS 1 AND LAGEOS 2 DATA FOR THE PERIOD 1984 – 2011

Yavor Chapanov and Ivan Georgiev

National Institute of Geophysics, Geodesy and Geography, BAS
Acad. G. Bonchev Str. Bl.3, Sofia 1113, Bulgaria
E-mail: yavor.chapanov@gmail.com

The laser ranging data of geodynamic satellites Lageos 1 from period April 1984 - December 2011 and Lageos 2 from period January 1993 - December 2011 are processed and analyzed. The processing of the observations is made by the latest version 4.2 of the program SLRP (Satellite Laser Ranging Processor), developed in the National Institute of Geophysics, Geodesy and Geography (NIGGG) at the Bulgarian Academy of Sciences (BAS). The SLR solution contains coordinates and velocities of the laser stations, Earth Orientation Parameters (EOP) and estimates of some Earth's dynamic and geometric parameters, such as: geogravitational constant; part of the Earth's geopotential coefficients; Love and Shida numbers; amplitudes and phases of the ocean tides; parameters connected with the satellites motion and laser observations modeling. The paper analyzes some of the important time series connected with the Earth orientation parameters (polar motion and Universal Time UT1) and gravity variations (the product of gravity constant G and Earth mass M, second harmonic J2 and third harmonic J3 of the Earth gravity potential), applicable in Geodetic and Geophysical research.

X Serbian-Bulgarian Astronomical Conference (X SBAC)
May 30 - June 3, 2016, Belgrade, Serbia
Book of Abstracts, Eds. M. S. Dimitrijević and M. K. Tsvetkov
Astronomical Observatory, Belgrade, 2016

Short talk

NONSINGULAR BIG BANG IN NONLOCAL MODIFIED GRAVITY

Ivan Dimitrijević

Mathematical Faculty, Studentski trg 16, 11001 Belgrade, Serbia
E-mail: ivand@matf.bg.ac.rs

After discovery of accelerating expansion of the Universe, there has been a renewed interest in gravity modification. One of promising approaches is nonlocal modification with the scalar curvature R in the action replaced by a suitable function $F(R,\Box)$, where $\Box$ is D'Alembert (Laplace-Beltrami) operator. In particular we analyze the modification in the form

$$S = \int \left(\frac{R - 2\Lambda}{16\pi G} + R^p F(\Box) R^q \right) \sqrt{-g} d^4 x$$

where is $F(\Box)$ an analytic function. We present a few a(t) nonsingular bounce cosmological solutions for the above two actions using FLRW metric.

X Serbian-Bulgarian Astronomical Conference (X SBAC)
May 30 - June 3, 2016, Belgrade, Serbia
Book of Abstracts, Eds. M. S. Dimitrijević and M. K. Tsvetkov
Astronomical Observatory, Belgrade, 2016

Short talk

ASTRONOMICAL CONCERNS AND THEIR SOLVING IN CONNECTION WITH THE ARCHEO-ASTRONOMICAL LOCALITY LEPENSKI VIR

Vladimir Đošović, Bojan Tajhmeister and Kristina Pantelić

Society of Astronomers of Serbia, Volgina 7, 11000 Belgrade, Serbia
E-mail: vladimirdjosowic@gmail.com

Archaeological site of Lepenski Vir is located on the bank of the river Danube in eastern Serbia. Lepenski Vir is a Mesolithic site (Srejović, 1969) which has been dated from 9500 BC to 5500 BC (Borić, Dimitrijević, 2007). The aim of this paper is testing the hypothesis in the paper of Ljubinka Babović where she assumes that the movement of the Sun has influenced the construction of houses on Lepenski Vir (Babovic, 2006). Description of the work and the first references were given by profesor Stevo Segan on the subject of "Processing of Astronomical Observations".

The sunrise was observed from the site, which appeared above the hills Treskavica across the river in Romania (Bajić, Pavlović, 2015). Babović has given the assumption that the positions of the houses are in correlation with sunrise. Also, the range of angles of inclination of the houses to the East corresponds to twice time bigger slope of the ecliptic relative to the equator, respectively that corresponds to the range of declination of the Sun.

Lepenski Vir site has been put into archaeological context. Movement and ephemeris of planets, the Sun and the Moon have been numerically modelled in C++ programing language (Morrison, Stephenson, 2004) The reconstruction of the sky during this period has been modelled. Place of the sunrise has been photographed *in situ*.

References

Babović, Ljubinka: 2006, Lepenski vir – the spatial annual calendar of the motion of the Sun Deity, *Publ. Astronomical Observatory Belgrade,* **80**, 337-342.

Bajić, Aleksandra; Pavlović, Hristivoje: 2015, Sunce Lepenskog Vira, Društvo „Vlašići", Beograd

Borić, Dušan; Dimitrijević, Vesna: 2007, Apsolutna hronologija i stratigrafija Lepenskog Vira, *Starinar,* **LVII**, 9-55.

Morrison, L. V., Stephenson, F. R.: 2004, Historical values of the Earth's clock error ΔT and the calculation of eclipses, *J. History Astron.,* **35(3)**, 327–336.

Srejović, Dragoslav: 1969, *Lepenski Vir: A New Prehistoric Culture in the Danube Valley,* Belgrade, SKZ.

X Serbian-Bulgarian Astronomical Conference (X SBAC)
May 30 - June 3, 2016, Belgrade, Serbia
Book of Abstracts, Eds. M. S. Dimitrijević and M. K. Tsvetkov
Astronomical Observatory, Belgrade, 2016

Short talk

DENSITY DISTRIBUTION FUNCTION IN A SELF-GRAVITATING INCOMPRESSIBLE TURBULENT FLUID

Sava Donkov and Ivan Stefanov

Department of Applied Physics, Technical University-Sofia ,
8 Kliment Ohridski Blvd, 1000 Sofia, Bulgaria
E-mail: savadd@tu-sofia.bg

We have set ourselves the task of obtaining the probability distribution function of the mass density of a self-gravitating incompressible turbulent fluid. We have applied a new approach that takes into account the fractal nature of the fluid. Using the medium equations we show that the total energy per unit mass is an invariant with respect to the fractal scales. As a next step we obtain a nonlinear differential equation for the probability distribution function of the mass density. It is solved numerically in several particular cases. For all considered values of the parameters the solutions look alike. They are power-law distributions with slope ranging between -1.5 and -2.

Short talk

SOFTWARE TOOL FOR TIME CONVERSION IN THE WFPDB

Nikolay Kirov[1], Milcho Tsvetkov[2] and Katya Tsvetkova[2]

[1]*New Bulgarian University, Montevideo 21, Sofia, Bulgaria*
[2]*Institute of Astronomy with NAO Rozhen, Bulgarian Academy of Sciences,*
72 Tsarigradsko Chausse Blvd., 1784 Sofia, Bulgaria
E-mail: nkkirov@gmail.com, milcho.tsvetkov@gmail.com,
katya.tsvetkova09@gmail.com

An important part of the digitization of astronomical photographic plates is the determination of the plate exposure beginning. The time of the observation, which can be found written in the astronomical logbooks or on the plate envelopes and even on the plates, is often in local time, or sidereal time, or in Julian dates. Timetool presented here is open source software (github.com/nkirov/timetool), which transforms any time format to Universal time. The tool reads the input data in the Wide-Field Plate Database (WFPDB - www.wfpdb.org) format and outputs the data in the same format, following some strong conventions about the file names.

X Serbian-Bulgarian Astronomical Conference (X SBAC)
May 30 - June 3, 2016, Belgrade, Serbia
Book of Abstracts, Eds. M. S. Dimitrijević and M. K. Tsvetkov
Astronomical Observatory, Belgrade, 2016

Short talk

GAUSSIAN PROCESSES FOR LONG-TERM OBSERVED CONTINUUM AND BROAD LINE LIGHT CURVES OF E1821+643

Andjelka Kovačević[1], Luka Č. Popović[2], Alla I. Shapovalova[3] and Dragana Ilić[1]

[1]*Faculty of Mathematics, University of Belgrade, Studentski trg 16, 11000 Belgrade, Serbia*
[2]*Astronomical observatory Belgrade, Volgina 7, 11160 Belgrade, Serbia*
[3]*Special Astrophysical Observatory of the Russian AS, Nizhnij Arkhyz, Karachaevo-Cherkasia, 369167, Russia*
E-mail: andjelka@matf.bg.ac.rs

A binary black hole candidate QSO E1821+643 has been monitored in the period from 1990 to 2014. Our findings suggest a possible periodicity in the continuum and broad emission line light curves (Shapovalova et al. 2016), which may be caused by orbital motion. To explore the periodicity indications we applied non-parametric models, i.e. the Gaussian processes, to the continuum and broad emission line light curves of QSO E1821+643.

References

Shapovalova, A. I., Popović, L. Č., Chavushyan, V. H., Burenkov, A. N., Ilić, D., Kollatschny, W., Kovačević, A. et al.: 2016, *ApJS*, accepted.

X Serbian-Bulgarian Astronomical Conference (X SBAC)
May 30 - June 3, 2016, Belgrade, Serbia
Book of Abstracts, Eds. M. S. Dimitrijević and M. K. Tsvetkov
Astronomical Observatory, Belgrade, 2016

Short talk

WZ SGE TYPE CATACLYSMIC VARIABLE ASASSN-14CL: SUPERHUMPS AND FLICKERING IN 2014 SUPEROUTBURST

G. Latev[1], V. Popov[2], S. Boeva[1], P. Nikolov[1] and B. Spassov[1]

[1]*Institute of Astronomy with NAO Rozhen, Bulgarian Academy of Sciences, 72 Tsarigradsko Chausse Blvd., 1784 Sofia, Bulgaria*
[2]*Department of Physics, Shumen University, 115, Universitetska Str., 1712 Shumen, Bulgaria*
E-mail: glatev@stro.bas.bg

We present simultaneous multicolour observations of the flickering of the WZ Sge type cataclysmic variable ASASSN-14cl after the superoutburst decline in August 2014. Using AAVSO data we obtain the average superhumps period Tsh = 0.059874 d (1h 26 min), the period evolution on the O-C diagram, and the range of V-band amplitude ΔA = 0.1-0.3 mag.

Short talk

VENUS IONOSPHERE ELECTRON PROPERTIES – CASSINI QUASI-THERMAL NOISE MEASUREMENTS

Mihailo M. Martinović[1,2,3], Arnaud Zaslavsky[1], Milan Maksimović[1] and Stevo Šegan[2]

[1]*LESIA, Observatoire de Paris, UPMC, Université Paris Diderot, CNRS, Meudon, France,*
[2]*Department of Astronomy, Faculty of Mathematics, Belgrade, Serbia*
[3]*IHIS Techno-experts d.o.o. - Research and Development Center, Belgrade, Serbia*
E-mail: mihailo.martinovic@obspm.fr

Quasi-thermal noise (QTN) spectroscopy is an accurate technique for in situ measurements of electron density and temperature in space plasmas. The QTN spectrum has a characteristic noise peak just above the plasma frequency produced by electron quasi-thermal fluctuations. This fact allows very accurate measurements of the electron density, while kinetic temperature of the plasma can be evaluated from the level of a power spectrum. In this work, we were able to deduce these plasma parameters during the first CASSINI flyby of Venus, since the thermal noise peak was visible by CASSINI/RPWS instrument on the closest approach (up to 284 km above the surface – deep in the ionosphere of the planet).

X Serbian-Bulgarian Astronomical Conference (X SBAC)
May 30 - June 3, 2016, Belgrade, Serbia
Book of Abstracts, Eds. M. S. Dimitrijević and M. K. Tsvetkov
Astronomical Observatory, Belgrade, 2016

Short talk

ASYMPTOTIC SOLUTION FOR SPATIALLY OPEN UNIVERSE WITH MATTER-DOMINATED EVOLUTION

Žarko Mijajlović[1] and Nadežda Pejović[2]

[1]*State Univ. of Novi Pazar, Novi Pazar, Serbia*
[2]*Faculty of Mathematics, Univ. of Belgrade, Serbia*
E-mail: zarkom@matf.bg.ac.rs, nada@matf.bg.ac.rs

In Mijajlović et al. (2012), we applied the theory of regularly varying functions in the asymptotic analysis of cosmological parameters for the expanding universe. This analysis was based on the Friedmann equations, particularly on the so called *acceleration equation*. We have shown there that this equation can be put in the form

$$\ddot{a}(t) + \frac{\mu(t)}{t^2} a(t) = 0, \tag{1}$$

where $\mu(t)$ explicitly depends on the other two fundamental cosmological parameters, the energy density $\rho = \rho(t)$ and the pressure of the material in the universe $p = p(t)$. If the matter-dominated evolution of the flat universe is assumed, even for the non-zero cosmological constant Λ, we proved there that the expression $\mu(t)$ depends solely on the density parameter $\Omega(t)$:

$$\mu(t) = \frac{2}{9} \cdot \frac{3\Omega - 2}{1 - \Omega} \left(\ln \left(\frac{1 + \sqrt{1 - \Omega}}{\sqrt{\Omega}} \right) \right)^2 . \tag{2}$$

Derivation of this formula is based on the formula for the age of the spatially flat universe with the cosmological constant Λ, due to Carroll at al. (1992), see also Liddle and Lyth (2000). In this paper we derive a similar formula to (2) for spatially open universe with matter-dominated evolution.

References

Carroll, S. M., Press, W. H., Turner, E. L.: 1992, *Ann. Rev. Astron. Astrophys*, **30**, 499.

Liddle, A. R., Lyth, L. H.: 2000, *Cosmological Inflation and Large-Scale Structure*, 400, Cambridge Univ. Press.

Mijajlović, Ž., Pejović, N., Šegan, S., Damljanović, G.: 2012, On asymptotic solutions of Friedmann equations, *Applied Mathematics and Computation*, **219**, 1273.

X Serbian-Bulgarian Astronomical Conference (X SBAC)
May 30 - June 3, 2016, Belgrade, Serbia
Book of Abstracts, Eds. M. S. Dimitrijević and M. K. Tsvetkov
Astronomical Observatory, Belgrade, 2016

Short talk

FLUCTUATING GOVERNING PARAMETERS IN GALAXY DYNAMO

E. A. Mikhailov and V. V. Pushkarev

Moscow State University, Faculty of Physics (Russia, 119991, Leninskie gory, bldg.1, str.2, MSU, Faculty of Physics)
E-mail: ea.mikhajlov@physics.msu.ru

Evolution of galaxy magnetic fields is described by the so-called dynamo theory. The equations include governing parameters that describe alpha-effect, differential rotation and turbulent diffusion. Usually these coefficients are connected with average kinematic characteristics of the interstellar medium. This approach is possible for "calm" galaxies. However, if there is intensive star formation, supernova explosions and other active processes, the parameters of the interstellar gas change. The distribution of the regions where the velocities of the interstellar medium are different is quite random. So it is useful to use the model of the galaxy dynamo with random coefficients (Moss et al 2015). Previous papers described fluctuating alpha-effect (Mikhailov, Modyaev 2015), but it is necessary to take into account the turbulent diffusion, too.

We use so-called no-*z* approximation to study the magnetic field (Moss 1995). It takes into account that the galaxy disc is thin, so the *z*-derivatives can be changed by algebraic expressions. We describe the fluctuations of alpha-effect and turbulent diffusion coefficients. Both asymptotic estimates of the field growth rate and the numerical simulations for such model have been made.

We have shown that the magnetic field can be changed a lot even for quite small fluctuations of the dynamo governing parameters. For example, the star formation that is 5-7 times higher than in the Milky Way can destroy the regular magnetic field.

References

Moss, D., Stepanov, R., Krause, M., Beck, R., Sokoloff, D.: 2015, *A&A*, **578,** A94.
Mikhailov, E. A., Modyaev, I. I.: 2015, *Baltic Astronomy,* **24**, 2, 194.
Moss, D.: 1995, *MNRAS,* **275**, 1, 191.

X Serbian-Bulgarian Astronomical Conference (X SBAC)
May 30 - June 3, 2016, Belgrade, Serbia
Book of Abstracts, Eds. M. S. Dimitrijević and M. K. Tsvetkov
Astronomical Observatory, Belgrade, 2016

Short talk

OPTICAL VARIABILITY OF DIFFERENT TYPES AGNs OBSERVED FROM ROZHEN OBSERVATORY

Rosa Victoria Munoz Dimitrova and Rumen Bachev

Institute of astronomy and NAO Rozhen,
Sofia 1784, Blvd. Tzarigradsko shaussee 72, Bulgaria
E-mail: rdimitrova@astro.bas.bg, bachevr@astro.bas.bg

Active galactic nuclei are the most luminous and variable objects in the Universe. It is still not known enough about the nature and characteristics of the optical variability of this kind of objects and it is important to monitor different types of AGNs to find more information about the accretion disc structures and properties, and respectively information of the central engine of the system.

We present our first results of long slit spectroscopic monitoring of different types Active Galactic Nuclei.

The data were taken with 2m RCC Telescope of Rozhen Observatory in 2015 for ~ 140h with specific light-reciever called FoReRo2 (two-channel Focal-Reducer Rozhen).

The result includes observations, reductions and light curves of broad emission lines and analysis of the data.

We are still selecting objects for future reverberation mapping campaigns.

X Serbian-Bulgarian Astronomical Conference (X SBAC)
May 30 - June 3, 2016, Belgrade, Serbia
Book of Abstracts, Eds. M. S. Dimitrijević and M. K. Tsvetkov
Astronomical Observatory, Belgrade, 2016

Short talk

COMETS – FIERY SWORD CONNECTION: A KRUSEDOL CLUE

Ретко Nedialkov[1] and Goran Andrejevic[2]

[1]*Department of Astronomy, University of Sofia, Sofia 1164*
[2]*Library of the Primary School "Vozd Karadjordje", Leskovac 16000*
E-mail: japet@phys.uni-sofia.bg

A half length image of St. Michael the Archangel holding a fiery sword above his head can be spotted during the initial frames of the Aleksandar Petrovic 1967 cult movie Skupljaci perja (Internet Movie Database). It turns out that the mural painting resides in the monastery of Krusedol in Frushka gora, Srem District of Serbia. The fresco is positioned on the west wall, just on the right side as one steps into the inner narthex of the church. The interior was covered with murals around 1750 by the painter Jov Vasiljevic and his fellows.

The first attempt to date the Bible with a transient event like a comet was done by Gunnar Norling (1953) who proposed a probable link between the mentioning an angel with a fiery sword and the appearance of the Halley's comet. Based on the iconographic material from the Bulgarian and Italian Renaissance Nedialkov & Enikova (2016 and references therein) reviewed all the astronomical, historical and ethnographical evidences that such a connection does really exist. In this report we address the possible connection between the depiction of the Archangel Michael with a fiery sword in the narthex of the Krusedol monastery and the passages of two bright comets in 1737 and 1744 (Hasegawa 1980).

Authors are thankful to Nenad Glisic, Branko Stevanovic and Biljana Begovic who provided them with information about the Krusedol mural painting.

References

Hasegawa, I.: 1980, *Vistas in Astronomy*, **24**, 59.

Internet Movie Database: 2016, *http://www.imdb.com/title/tt0062277.*

Nedialkov, P., Enikova, P.: 2016, *40 years jubilee of the Smolyan planetarium, Bulgaria (conference talk).*

Norling, G.: 1953, *Actes du VII Congres International d'Histoire des Sciences: Jerusalem*, **8**, 458.

X Serbian-Bulgarian Astronomical Conference (X SBAC)
May 30 - June 3, 2016, Belgrade, Serbia
Book of Abstracts, Eds. M. S. Dimitrijević and M. K. Tsvetkov
Astronomical Observatory, Belgrade, 2016

Short talk

POLARIZATION IN AGN BROAD EMISSION LINES – THE CENTRAL SOURCE ANIZOTROPY AND GAS KINEMATICS

Djordje V. Savić[1], Rene W. Goosmann[2], Martin Gaskell[3], Frederic Marin[2] and Luka Č. Popović[1]

[1]*Astronomical Observatory Belgrade, Volgina 7, 11060 Belgrade, Serbia*
[2]*Observatoire Astronomique de Strasbourg, Université de Strasbourg, CNRS, UMR 7550, 11 rue de l'Université, 67000 Strasbourg, France*
[3]*University of California, Santa Cruz, 1156 High St, Santa Cruz, CA 95064, United States*
E-mail: djsavic@aob.rs

We investigate the polarized broad line (BLR) emission in active galactic nuclei (AGN) using the 3D Monte Carlo radiative transfer code STOKES (Goosmann et al. 2013). We consider a model where the central engine consists of two point-like continuum sources, with one of the continuum sources being off-centered. The BLR is modeled using a flared-disk geometry with Keplerian rotation, where absorption, re-emission and scattering can occur. We discuss our results in the frame of the close super-massive black hole binary hypothesis.

References

Goosmann, R. W., Gaskell, C. M., Marin, F.: 2013, *Advances in Space Research,* **54**, 1341.

X Serbian-Bulgarian Astronomical Conference (X SBAC)
May 30 - June 3, 2016, Belgrade, Serbia
Book of Abstracts, Eds. M. S. Dimitrijević and M. K. Tsvetkov
Astronomical Observatory, Belgrade, 2016

Short talk

SIMPLIFIED MODEL OF LINE PROFILE VARIABILITY FROM ECCENTRIC ORBITS OF SUPERMASSIVE BINARY BLACK HOLE SYSTEMS

Marijana Smailagić and Edi Bon

Astronomical Observatory, Volgina 7, 11060 Belgrade Serbia
E-mail: marijana@aob.rst

One of possible explanations for variability of active galactic nuclei involves existence of a system of sub-parsec supermassive binary black holes (SMBBHs) in their centre. SMBBHs are expected to exist as a result of major mergers in the Universe, and some SMBBH systems with projected separation of about few parsecs were directly observed. In this work we model the emission line profiles of active galactic nuclei assuming that a SMBBH system is located in its core, and that the accreting gas inside the circumbinary disk is photo ionized by mini accretion disk emission around each SMBBH. We calculate variations of emission line properties for different eccentricities and orientations of SMBBHs' orbits and mass ratios of black holes. We examine how these parameters influence time variability in total line flux and line shifts.

X Serbian-Bulgarian Astronomical Conference (X SBAC)
May 30 - June 3, 2016, Belgrade, Serbia
Book of Abstracts, Eds. M. S. Dimitrijević and M. K. Tsvetkov
Astronomical Observatory, Belgrade, 2016

Short talk

РАДИОАСТРОНОМИЯ В ЛАТВИИ – ИСТОРИЯ И ПЕРСПЕКТИВЫ РАЗВИТИЯ

Ивар Шмелд

[1]*Ventspils University College, VIRAC Inzenieru 101,*
Ventspils LV-3600 Latvia
E-mail: ivarss@venta.lv

Дается обзор развития радиоастрономии в Латвию Первым этапом развития являются работы по проектированию и попытки построения интерферометра по схеме креста Миллса в 50 – 60гг прошлого века – недалеко от местечка Балдоне примерно 30 км от Риги. Параллельно развивалась и солнечная радиоастрономия, которая берет свое начало с небольших радиотелескопов, построенных на основе передвижных радарных станции времени второй мировой войны, а в начале 70-тых вступил в строй 10 метровый радиотелескоп дециметрового диапазона. Исследования Солнца с помощью этого телескопа продолжались вплоть до 1993 года.

Начиная с 1994 АН Латвии от уходящей советской армии получила 32 метровый радиотелескоп недалеко от Вентспилса. Сейчас этот телескоп является основой Вентспилского Радиоастрономического центра Вентспилской Высшей школы. В сообщение дальше дается обзор истории развития Центра, его теперешнего состояния и перспектив развития.

X Serbian-Bulgarian Astronomical Conference (X SBAC)
May 30 - June 3, 2016, Belgrade, Serbia
Book of Abstracts, Eds. M. S. Dimitrijević and M. K. Tsvetkov
Astronomical Observatory, Belgrade, 2016

Short talk

DENSITY SCALING RELATION IN ORION A: EFFECTS OF REGION SELECTION

Orlin Stanchev[1], Todor Veltchev[1] and Sava Donkov[2]

[1]*Faculty of Physics,Department of Astronomy, University of Sofia, 5James Bourchier Blvd., Sofia 1164, Bulgaria*
[2]*Department of Applied Physics, Technical University-Sofia, 8 Kliment Ohridski Blvd, 1000 Sofia, Bulgaria*
E-mail: o_stanchev@phys.uni-sofia.bg

Recenlty Stanchev et al. (2015) proposed a method to derive density scaling relation in a molecular cloud from analysis of the probability distribution function of column density. A possible bias of the procedure is its dependence on the selection of probe regions of different size and location in the considered cloud. We present a study of this issue, using a PLANCK map (dust opacity) of Orion A.

Short talk

MULTI-WAVELENGTH OBSERVATIONS OF A TWO-RIBBON SOLAR FLARE CAUSED BY FILAMENT ERUPTION

Tsvetan Tsvetkov, Kostadinka Koleva, Peter Duchlev and Momchil Dechev

Institute of Astronomy with National Astronomical Observatory, 72, Tsarigradsko Chaussee Blvd., 1784, Sofia, Bulgaria
E-mail: tstsvetkov@astro.bas.bg

A study of a two-ribbon flare preceded by a filament/prominence eruption is presented. The event was observed between 00:00 and 08:00 UT on 2014 February 18 in a quiet region between NOAA active regions 11982 and 11977. The multi-wavelength analysis of the eruptive filament and following two-ribbon flare was made using data obtained from the Solar Dynamics Observatory (SDO), Solar Terrestrial Relations Observatory (STEREO) B and ground-based observatories. The kinematics and morphology, as well as the evolution of the overlying coronal fields were examined using the observations in optical, EUV and radio wavelengths. The trigger mechanisms of the eruptions were discussed.

POSTER PAPERS

X Serbian-Bulgarian Astronomical Conference (X SBAC)
May 30 - June 3, 2016, Belgrade, Serbia
Book of Abstracts, Eds. M. S. Dimitrijević and M. K. Tsvetkov
Astronomical Observatory, Belgrade, 2016

Poster paper

SELECTED SERBIAN ASTRONOMERS IN THE WEB OF SCIENCE IN THE PERIOD 1977-2014

Bojan Arbutina

Faculty of Mathematics, University of Belgrade, Studentski trg 16, 11000 Belgrade, Serbia
E-mail: bojan.arbutina@gmail.com

I discuss the number of papers and citation information for selected Serbian astronomers in the Thomson Reuters' Web of Science for the period 1977-2014.

X Serbian-Bulgarian Astronomical Conference (X SBAC)
May 30 - June 3, 2016, Belgrade, Serbia
Book of Abstracts, Eds. M. S. Dimitrijević and M. K. Tsvetkov
Astronomical Observatory, Belgrade, 2016

Poster paper

ON THE ANOMALOUS LOW SPONTANEOUS EMISSION RATES FOR p-SERIES OF SODIUM DUE TO THE EFFECT OF NATURAL FÖRSTER RESONANCE

K. N. Arefieff[1], N. N. Bezuglov[1], M. S. Dimitrijević[2,3,4], A. N. Klycharev[1], A. A. Mihajlov[5] and V. A. Srećković[5]

[1]*St. Petersburg State University, St. Petersburg, 198904 Russia*
[2]*Astronomical Observatory, Volgina 7, 11160 Belgrade 74, Serbia*
[3]*IHIS-Technoexperts, Bežanijska 23, 11080 Zemun, Serbia*
[4]*Observatoire de Paris, 92195 Meudon Cedex, France*
[5]*University of Belgrade, Institute of Physics, P. O. Box 57, 11001, Serbia*
E-mail: vlada@ipb.ac.rs

In our work (Arefieff *et al.* 2015) we paid attention to a strong blockade of dipole matrix elements for l'- and l-series in the vicinity of Föster resonance. The latter occurs when the diference $\Delta\delta=\delta_{l'}-\delta_l$ between the series quantum defects becomes equal to 0.5. Table 1 demonstrates that the case of s-, p-series for Na atom practically meats this requirement. Radiative rate constants $A_{nl}=1/\tau_{nl}$ of Rydberg states spontaneous decay were shown in (Bezuglov *et al.* 1991) to obey the power law $A_{nl}=\alpha\bullet C_l\, n^{*-3}/(l+0.5)$ regarding to the effective quantum number $n^*=n-\delta_l$. The constant $\alpha=1.18\cdot10^{10}s^{-1}$ while C_p-coefficients are presented in Table 2 that demonstrates the anomalous law value (in bold) for Na case.

Table 1 Quantum defect δ_l for s-, p-series of the alkali and hydrogen atoms

l	Li	Na	K	Rb	Cs	H
s	0.40	1.35	2.19	3.13	4.06	0
p	0.04	0.85	1.71	2.66	3.59	0

Table 2 C_l-coefficient for s-, p-series of the alkali and hydrogen atoms

l	Li	Na	K	Rb	Cs	H
s	0.025	0.015	0.017	0.017	0.018	0.013
p	0.069	**0.014**	0.051	0.075	0.061	1.00

References

Arefieff, K. N., Bezuglov, N. N., *et al*: 2015, *J. Astrophys. Astr.*, **36**, 613.
Bezuglov, N. N., Borisov, E. N., *et al.*: 1991, *Sov. Phys. Usp*, **34**, 3.

X Serbian-Bulgarian Astronomical Conference (X SBAC)
May 30 - June 3, 2016, Belgrade, Serbia
Book of Abstracts, Eds. M. S. Dimitrijević and M. K. Tsvetkov
Astronomical Observatory, Belgrade, 2016

Poster paper

LIGHT CURVE SOLUTIONS OF FOUR DETACHED ECLIPSING *KEPLER* BINARIES

Teodora Atanasova and Diana Kjurkchieva

Department of Physics, University of Shumen, 115 Universitetska Str., Shumen, Bulgaria
E-mail: t.atanasova@shu.bg, d.kyurkchieva@shu-bg.net

We carried out light curve solutions of four detached binaries with circular orbits, observed by *Kepler* (KIC 5285607, KIC 5080652, KIC 11975363, KIC 9236858). As a result the orbital inclinations, temperatures and relative radii of their components were determined. We found additional short-term and long-term variability of the targets superposed on the eclipses.

X Serbian-Bulgarian Astronomical Conference (X SBAC)
May 30 - June 3, 2016, Belgrade, Serbia
Book of Abstracts, Eds. M. S. Dimitrijević and M. K. Tsvetkov
Astronomical Observatory, Belgrade, 2016

Poster paper

INFLUENCE OF SOLAR X-RAY FLARES ON RADIO SIGNAL PROPAGATION IN THE LOW IONOSPHERE

Jovan Bajčetić[1] and Aleksandra Nina[2]

[1]Department of Telecommunications and Information Science, University of Defence, Military Academy, Belgrade, Serbia
[2]Institute of Physics, University of Belgrade, Belgrade, Serbia
E-mail: bajce05@gmail.com

In addition to pure scientific significance, and possible applications in prediction of elementary disasters, ionospheric investigation can be of practical use in the field of radio communications. Namely, radio wave propagation over long distances is enabled by multiple reflections of electromagnetic (EM) waves inside the Earth - ionosphere waveguide and is dependent of ionospheric properties which can be changed by X-flares, Lyman-α radiation, lightning, geomagnetic storms and many other effects. Also, ionospheric plasma changes characteristics of satellite signals.

In this paper, we study general characteristics of a solar X-ray flare (the most important sudden perturber of the low ionosphere) influence on telecommunication radio signals. We based our analysis on the low ionospheric monitoring using VLF signal while the D-region plasma parameters and signal propagation characteristics are calculated using procedures explained in Grubor et al. (2008), Nina and Čadež (2014) and Bajčetić et al. in press.

References

Bajčetić J., Nina A., Čadež V.M., Todorović B. M.: *Thermal Science*, in press, doi:10.2298/TSCI141223084B.
Grubor, D. P., Šulić, D. M., Žigman, V.: 2008, *Annales Geopysicae*, **26**, 1731.
Nina A., Čadež V.M.: 2014, *Advances in Space Research*, **54**(7), 1276.

X Serbian-Bulgarian Astronomical Conference (X SBAC)
May 30 - June 3, 2016, Belgrade, Serbia
Book of Abstracts, Eds. M. S. Dimitrijević and M. K. Tsvetkov
Astronomical Observatory, Belgrade, 2016

Poster paper

A SIMILARITY BETWEEN THE SOLAR SYSTEM AND THE KEPLER'S THREE-PLANETARY EXTRASOLAR SYSTEMS

Borislav Stanishev Borisov

University of Shumen, Faculty of Natural Sciences,
115, Universitetska Str., 9712 Shumen, Bulgaria
E-mail: b.st.borisov@abv.bget

All known 3-planetary systems discovered by Kepler are included in this study. A linear transformation of planetary distances for each system is made so the distance to the innermost planet is equal to zero and the distance to the outermost one is equal to one. A distribution histogram of the transformed distance to the third planet is built. It is fitted with two Gaussian functions. Based on this fit the Earth-Sun distance is derived as it has never been included in the calculations. The accuracy of the result is more than 99.5%. All this shows that there is a simple linear relationship between the planetary distances in the Solar system and the planetary distances in most extrasolar systems.

X Serbian-Bulgarian Astronomical Conference (X SBAC)
May 30 - June 3, 2016, Belgrade, Serbia
Book of Abstracts, Eds. M. S. Dimitrijević and M. K. Tsvetkov
Astronomical Observatory, Belgrade, 2016

Poster paper

SOLAR INFLUENCE ON SUBDECADAL VARIATIONS OF EARTH ROTATION, MSL AND CLIMATE, DERIVED BY HARMONICS OF HALE, JOSE, DE VRIES AND SUESS CYCLES

Yavor Chapanov[1], Cyril Ron[2] and Jan Vondrak[2]

[1]*National Institute of Geophysics, Geodesy and Geography, BAS*
Acad. G. Bonchev Str. Bl.3, Sofia 1113, Bulgaria
[2]*Astronomical Institute, Academy of Sciences of Czech Republic;*
Bočni II 1401, 141 00 Prague 4, Czech Republic
E-mail: yavor.chapanov@gmail.com

The solar activity cycles affect all surface geosystems, including weather and climate indices, winds, rains, snow covers, mean sea level, river streamflows and other hydrological cycles. The mean sea level and polar ice changes cause common variations of the principal moments of inertia and Earth rotation with decadal, centennial and millennial periods. The mean sea level, Earth rotation and climate indices have also some oscillations with periods from several months to 40 years, whose origin is not connected with the known tidal, seasonal or solar effects. The shape of solar cycles is rather different from sinusoidal form, so they affect geosystems by many short-term harmonics. A possible solar origin of subdecadal variations of Earth rotation, mean sea level and climate indices is investigated by the harmonics of 22-year Hale cycle and 45-year cycle of equatorial solar asymmetry. The solar influence on 13-year oscillations of climate and Earth rotation is analyzed by the harmonics of Jose, de Vries and Suess cycles with centennial periods of 178.7, 208 and 231 years.

X Serbian-Bulgarian Astronomical Conference (X SBAC)
May 30 - June 3, 2016, Belgrade, Serbia
Book of Abstracts, Eds. M. S. Dimitrijević and M. K. Tsvetkov
Astronomical Observatory, Belgrade, 2016

Poster paper

WEBT PROJECT AND OBSERVATIONS AT THE 60 CM ASV TELESCOPE

Goran Damljanovic[1], Oliver Vince[1], Rumen Bachev[2] and Werner W. Zeilinger[3]

[1]*Astronomical Observatory, Volgina 7, 11060 Belgrade, Serbia*
[2]*Institute of Astronomy with NAO, Bulgarian Academy of Sciences, Tsarigradsko Shosse Blvd. 72, 1784 Sofia, Bulgaria*
[3]*Department of Astrophysics, University of Vienna, Wien, Austria*
E-mail: gdamljanovic@aob.rs

At the Astronomical Observatory (AOB, Belgrade, Serbia) we started the observations of blazars using the 60 cm telescope at ASV (Astronomical Station Vidojevica) at the beginning of 2013. Some time after that, we established a cooperation with the Astronomical Observatory of Torino (Osservatorio Astrofisico di Torino) as an international center for the Whole Earth Blazar Telescope – WEBT program. Also, we started the cooperation with colleagues from Bulgaria and Austria about WEBT project, and used few Bulgarian telescopes (at the Belogradchik AO and Rozhen Observatory) and the 1.5 m Leopold-Figl one of University of Vienna (Austria). As result, about 20 objects were observed until the end of 2015 (in BVRI filters). Few papers about these objects were published (Carnerero *et al.* 2015, Raiteri *et al.* 2015, Furniss *et al.* 2015, Bhatta *et al.* 2015). Here, we present some of our WEBT observations and results.

References

Carnerero, M. I., et al.: 2015, *MNRAS,* **450**, 2677.
Raiteri, C. M., et al.: 2015, *MNRAS,* **454**, 353.
Furniss, A., et al.: 2015, *Astrophysical Journal,* **812,** 65.
Bhatta, G., et al.: 2015, *Astrophysical Journal Letters,* **809,** L27.

X Serbian-Bulgarian Astronomical Conference (X SBAC)
May 30 - June 3, 2016, Belgrade, Serbia
Book of Abstracts, Eds. M. S. Dimitrijević and M. K. Tsvetkov
Astronomical Observatory, Belgrade, 2016

Poster paper

STARK BROADENING PARAMETERS OF S II SPECTRAL LINES FOR STELLAR PLASMA CONDITIONS

Milan S. Dimitrijević[1], Nenad Milovanović[1], Zoran Simić[1] and Sylvie Sahal-Bréchot[2]

[1]*Astronomical Observatory, Volgina 7, 11060 Belgrade, Serbia*
[2]*LERMA, Observatoire de Paris, PSL Research University, CNRS, Sorbonne Universities,UPMC Univ. Paris 06, France*
E-mail: mdimitrijevic@aob.rs, nmilovanovic@aob.rs, zsimic@aob.rs, sylvie.sahal-brechot@obspm.fr

Spectral lines of ionized sulfur have been observed in stellar spectra (see e.g. Adelman and Pintaldo (2000), Adelman et al. (2001)), so that their Stark broadening parameters are needed for sulfur abundance determination, radiative transfer calculations as well as for stellar spectra analysis and synthesis. Stark broadening is particularly significant for white dwarfs, post AGB stars, pre-white dwarfs and for A type stars. By using semiclassical perturbation method, we have determined widths and shifts for 12 spectral lines of ionized sulfur, broadened by collisions with electrons, protons and helium ions, for plasma conditions of interest for stellar atmospheres. The obtained results will be compared with published experimental and theoretical data and also included in the STARK-B database (http://stark-b.obspm.fr – Sahal-Bréchot et al., 2015), a part of Virtual Atomic and Molecular Data Center (VAMDC - http://www.vamdc.org – Dubernet et al., 2010).

References

Adelman, S. J., Pintaldo, O.I.: 2000, *Astron. Astrophys.*, **354**, 899.
Adelman, S. J. et al.: 2001, *Mon. Not. R. Astron. Soc.*, **328**, 1144.
Dubernet, M. L. et al.: 2010, *J. Quant. Spectrosc. Radiat. Transfer*, **111**, 2151.
Sahal-Bréchot, S., Dimitrijević, M. S., Moreau, N., Ben Nessib, N.: 2015, *Phys. Scripta*, **50**, 054008.

X Serbian-Bulgarian Astronomical Conference (X SBAC)
May 30 - June 3, 2016, Belgrade, Serbia
Book of Abstracts, Eds. M. S. Dimitrijević and M. K. Tsvetkov
Astronomical Observatory, Belgrade, 2016

Poster paper

ON THE INFLUENCE OF STARK BROADENING ON Xe VIII SPECTRAL LINES IN DO WHITE DWARFS

Milan S. Dimitrijević[1], Zoran Simić[1], Andjelka Kovačević[2] and Sylvie Sahal-Bréchot[3]

[1]*Astronomical Observatory, Volgina 7, 11060 Belgrade, Serbia*
[2]*Department for Astronomy, Faculty for Mathematics, Studentski Trg 16, 11000 Belgrade,Serbia*
[3]*LERMA, Observatoire de Paris, PSL Research University, CNRS, Sorbonne Universities,UPMC Univ. Paris 06, France*
E-mail: mdimitrijevic@aob.rs, zsimic@aob.rs, andjelka@matf.bg.ac.rs, sylvie.sahal-brechot@obspm.fr

Recently, Stark broadening parameters have been calculated for Xe VIII $6s\ ^2S_{1/2} - 6p\ ^2P^o_{1/2}$ and $7s\ ^2S_{1/2} - 7p\ ^2P^o_{1/2}$ transitions, for broadening by electron, proton, and He III impacts, using the semi-classical perturbation approach (Dimitrijević et al. 2015).

The obtained results have been used to study the influence of Stark broadening on spectral lines in DO white dwarf atmospheres and importance of Stark broadening due to collisions with electrons, protons and He III ions is shown.

References

Dimitrijević, M. S., Simić, Z., Kovačević, A., Valjarević, A., Sahal-Bréchot, S.: 2015, *MNRAS*, **454**, 1736.

X Serbian-Bulgarian Astronomical Conference (X SBAC)
May 30 - June 3, 2016, Belgrade, Serbia
Book of Abstracts, Eds. M. S. Dimitrijević and M. K. Tsvetkov
Astronomical Observatory, Belgrade, 2016

Poster paper

ON THE STARK BROADENING OF Si III SPECTRAL LINES IN STELLAR PLASMA

Milan S. Dimitrijević[1], Zoran Simić[1] and Sylvie Sahal-Bréchot[2]

[1]*Astronomical Observatory, Volgina 7, 11060 Belgrade, Serbia*
[2]*LERMA, Observatoire de Paris, PSL Research University, CNRS, Sorbonne Universities, UPMC Univ. Paris 06, France*
E-mail: mdimitrijevic@aob.rs, zsimic@aob.rs, sylvie.sahal-brechot@obspm.fr

Silicon atoms and ions in various ionisation stages are present in cosmic plasma and the Stark broadening of the their spectral lines is important for analysis, investigation and modelling of stellar plasma, especially for white dwarfs and hot stars of A and late B spectral type. They are also of interest for example for research and synthesis of stellar spectra, abundance determinations and radiative transfer calculations. In order to provide the needed Stark broadening parameters, calculations have been performed for a number of Si III spectral lines broadened by electron-, proton-, and helium ion-impacts, by using the semiclassical perturbation theory. The obtained results have been used to demonstrate the importance of Stark broadening in the spectra of A-type stars.

We note as well, that the obtained new Stark broadening data will be implemented in the STARK-B database (http://stark-b.obspm.fr – Sahal-Bréchot et al., 2015), a part of Virtual Atomic and Molecular Data Center (VAMDC - http://www.vamdc.org – Dubernet et al., 2010).

References

Dubernet, M. L. et al.: 2010, *J. Quant. Spectrosc. Radiat. Transfer*, **111**, 2151.

Sahal-Bréchot, S., Dimitrijević, M. S., Moreau, N., Ben Nessib, N.: 2015, *Phys. Scripta*, **50**, 054008.

X Serbian-Bulgarian Astronomical Conference (X SBAC)
May 30 - June 3, 2016, Belgrade, Serbia
Book of Abstracts, Eds. M. S. Dimitrijević and M. K. Tsvetkov
Astronomical Observatory, Belgrade, 2016

Poster paper

SHORT-TERM PERTURBATIONS IN HIGH AND MIDDLE LATITUDE LOW IONOSPHERE UNDER EUROPE INDUCED BY GRB$_S$

Ana Djulaković[1], Aleksandra Nina[2], Saša Simić[3], Vladimir A. Srećković[2] and Luka Č. Popović[4]

[1]*Faculty of Physics, University of Belgrade, Belgrade, Serbia*
[2]*Institute of Physics, University of Belgrade, Belgrade, Serbia*
[3]*Faculty of Science, Department of Physics, University of Kragujevac, Kragujevac, Serbia*
[4]*Astronomical Observatory of Belgrade, Belgrade, Serbia*
E-mail: ana_djulakovic@yahoo.com

In this study, we analyzed the low ionospheric short-term reactions connected with 54 gamma ray bursts (GRBs) registered in the energy range between 15 eV and 150 eV using Swift's Burst Alert Telescope device (BAT) in the period between 31 August 2009 and 25 November 2012. We based our analysis on the low ionospheric monitoring using the 37.5 kHz low frequency (LF) signal emitted by the NRK transmitter located in Grindavik (Iceland) and received at Institute of Physics in Belgrade (Serbia). The area through the signal from Iceland passes is characteristic because of high latitude location of the transmitter apropos a significant penetration of charged particles from outer space in the low ionosphere due to the curvature of the magnetic lines of force near the North Pole.

In addition to analysis of full sample (Nina *et al.* 2015) we study dependences of characteristics of GRBs and considered plasma medium on detectability LF signal short-term variations. The obtained results based on statistical analysis confirm detectability of short-term changes induced by GRBs in the analyzed ionospheric area. Significance of this study is for science investigations (geophysics, astrophysics) as well as in possible practical applications in telecommunication technologies.

References

Nina A., Simić S., Srećković V. A., Popović L. Č.: 2015, *Geophysical Research Letters,* **42** (19), 8250.

Poster paper

MULTICOLOUR PHOTOMETRIC STUDY OF T TAURI VARIABLES

Sunay Ibryamov[1,2], Evgeni Semkov[2] and Stoyanka Peneva[2]

[1]*Department of Theoretical and Applied Physics, University of Shumen, BG-9712 Shumen*
[2]*Institute of Astronomy and NAO, Bulgarian Academy of Sciences, BG-1784 Sofia*
E-mail: sibryamov@shu.bg

In this poster we present the results from a long-term multicolour CCD photometric observations of 22 pre-main sequence stars located in the dense molecular cloud L935, named 'Gulf of Mexico', in the field between the North America and Pelican nebulae. The long-term multicolour photometric observations of pre-main sequence stars are very important for their exact classification. The stars from our study exhibit different types of photometric variability in all optical passbands. Using our BVRI observations and data published by other authors, we tried to define the reasons for the observed brightness variations.

Poster paper

LONG-TERM UBVRI PHOTOMETRY OF THE PMS STAR V350 CEPHEI

Sunay Ibryamov[1,2], Evgeni Semkov[2] and Stoyanka Peneva[2]

[1]*Department of Theoretical and Applied Physics, University of Shumen, BG-9712 Shumen*
[2]*Institute of Astronomy and NAO, Bulgarian Academy of Sciences, BG-1784 Sofia*
E-mail: sibryamov@shu.bg

In this poster we present the results from UBVRI optical photometric observations of the pre-main sequence star V350 Cep collected during the period from 2004 to 2016. V350 Cep is located in the field of the reflection nebula NGC 7129, a region with active star formation. The star was discovered in 1977 due to its remarkable increase in brightness by more than 5 mag in R-band. In previous studies, the star was considered a to be a potential FUor or EXor eruptive variable. Our data suggest that during the period of observations the star maintains its maximum brightness with low amplitude photometric variations in all optical passbands.

X Serbian-Bulgarian Astronomical Conference (X SBAC)
May 30 - June 3, 2016, Belgrade, Serbia
Book of Abstracts, Eds. M. S. Dimitrijević and M. K. Tsvetkov
Astronomical Observatory, Belgrade, 2016

Poster paper

ACADEMICIANS MILUTIN MILANKOVIĆ AND VOJISLAV MIŠKOVIĆ – CORRESPONDENCE

Natalija Janc[1], Vojislava Protić Benišek[2], Vladimir Benišek[2], Milivoj B. Gavrilov[3], Luka Č. Popović[2] and Slobodan B. Marković[3]

[1]415 Old Trail Road, Baltimore, MD 21212, USA
[2]Astronomical Observatory, Volgina 7, 11060 Belgrade, Serbia
[3]University of Novi Sad, Faculty of Sciences, Laboratory for Paleoenvironmental Reconstruction, Trg Dositeja Obradovića 3, 21000 Novi Sad, Serbia
E-mail: natalijanc@earthlink.net

This paper presents a part of the unpublished correspondence between Milutin Milanković and Vojislav Mišković from the period 1924 to 1930. At the beginning of the correspondence, Mišković worked in the French observatory "Observatoire du Mont Gros" in Nice, while Milanković was in Belgrade. They discussed the PhD thesis of Mišković, his return to Bel-grade, getting him a position at the University, founding and building of the Astronomical Observatory in Belgrade, Milanković's research, publishing of his papers, his encounters with other scientists, his acceptance speech in the Serbian Academy of Science, and other topics, as well as the collabo-ration of Milanković with Alfred Wegener. With the help of Mišković, Mi-lanković wanted to better present Wegener's and his own research in French scientific journals. Milanković was also very interested for publications of the Astronomical Observatory in Belgrade, where he had found data that he could use in his research.

Mišković considered the letters from Milanković to be very important and mentioned them even in his obituary on occasion of Milanković's death. Proper care of those letters by Mišković himself, and later by his successors at the Astronomical Observatory in Belgrade, made it possible to have them now publicly available.

Analysis of their content gives additional insight into the life and work of these two Serbian scientists, especially about Milutin Milanković, whose contribution to the creation of a astronomical theory of Ice Ages has become an indispensable reading in the world of science, both in the past and in the present.

X Serbian-Bulgarian Astronomical Conference (X SBAC)
May 30 - June 3, 2016, Belgrade, Serbia
Book of Abstracts, Eds. M. S. Dimitrijević and M. K. Tsvetkov
Astronomical Observatory, Belgrade, 2016

Poster paper

MOL-D DATABASE FOR MOLECULAR COLLISIONAL AND RADIATIVE PROCESSES

Darko Jevremović[1], Veljko Vujčić[1], Anatolij A. Mihajlov[2], Vladimir A. Srećković[2], Ljubinko M. Ignjatović[2] and Milan S. Dimitrijević[1,3,4]

[1]*Astronomical Observatory, Volgina 7, 11160 Belgrade 74, Serbia*
[2]*University of Belgrade, Institute of Physics, P. O. Box 57, 11001, Serbia*
[3]*IHIS-Technoexperts, Bežanijska 23, 11080 Zemun, Serbia*
[4]*Observatoire de Paris, 92195 Meudon Cedex, France*
E-mail: vlada@ipb.ac.rs

We report the current stage of the MOL-D database and web service within the virtual atomic and molecular data center (VAMDC). It is devoted to modelling of stellar atmospheres, laboratory plasmas, technological plasmas etc. The data for hydrogen are for example important for elaboration of atmosphere models of solar and near solar type stars and for radiative transport consideration as well as for kinetic of stellar and other astrophysical plasmas (Mihajlov *et al.* 2007). The helium data are of interest particularly for helium-rich white dwarf atmospheres investigations. The initial stage of development was done at the end of 2015, since all the existing calculated data for photodissociation cross-sections of hydrogen H_2^+ and helium He_2^+ molecular ions and the corresponding averaged thermal photodissociation cross-sections have now been implemented (Vujčić *et al.* 2015). We are now beginning the second stage of the development of MOL-D. At the moment we are including new cross-section and rate coefficients data about processes which involve species such as HeH^+, LiH^+, NaH^+, SiH^+ which are important for the exploring of the interstellar medium as well as for the early Universe chemistry and for the modeling of stellar and solar atmospheres.

References

Mihajlov, A. A., Ignjatović, Lj. M., Sakan, N. M. Dimitrijević, M. S.: 2007, *Astronomy and Astrophysics,* **2**, 749.

Vujčić, V., Jevremović, D., Mihajlov, A. A., Ignjatović, Lj. M., Srećković, V. A., Dimitrijević, M. S., Malović, M.: 2015, *Journal of Astrophysics and Astronomy,* **4**, 22.

X Serbian-Bulgarian Astronomical Conference (X SBAC)
May 30 - June 3, 2016, Belgrade, Serbia
Book of Abstracts, Eds. M. S. Dimitrijević and M. K. Tsvetkov
Astronomical Observatory, Belgrade, 2016

Poster paper

OBSERVATIONS AND LIGHT CURVE SOLUTIONS OF THE ECLIPSING BINARIES USNO-B1.0 1395-0370184 AND USNO-B1.0 1395-0370731

Diana Kjurkchieva[1], Velimir Popov[2], Doroteya Vasileva[1] and Nikola Petrov[3]

[1] *Department of Physics, Shumen University, 115, Universitetska, 9700 Shumen, Bulgaria*
[2] *IRIDA Observatory, 57 Cherkovna str., Sofia, Bulgaria*
[3] *Institute of Astronomy and NAO, Bulgarian Academy of Sciences, Tsarigradsko shossee 72, Sofia, Bulgaria*
E-mail: d.kyurkchieva@shu-bg.net

The paper presents our photometric observations of two short-period eclipsing binaries. As a result of the light curve solutions we determined the orbital inclinations and temperatures and relative radii of their components as well as the parameters of cool spots reproducing the distortions of their light curves. We found that USNO-B1.0 1395-0370184 and USNO-B1.0 1395-0370731 are overcontact binaries with K components. Our photometric observations led to considerably improvement of their orbital periods.

X Serbian-Bulgarian Astronomical Conference (X SBAC)
May 30 - June 3, 2016, Belgrade, Serbia
Book of Abstracts, Eds. M. S. Dimitrijević and M. K. Tsvetkov
Astronomical Observatory, Belgrade, 2016

Poster paper

ON THE STARK BROADENING OF Co II SPECTRAL LINES

Zlatko Majlinger[1], Milan S. Dimitrijević[2,3,4] and Zoran Simić[2]

[1]*Faculty of Humanities and Social Sciences, University of Rijeka, 51000 Rijeka, Croatia*
[2]*Astronomical Observatory, Volgina 7, 11060 Belgrade, Serbia*
[3]*LERMA, Observatoire de Paris, UMR CNRS 8112, UPMC, 92195 Meudon Cedex, France,*
[4]*IHIS Techno-Experts, Batajnički put 23, 11080 Belgrade-Zemun, Serbia,*
E-mail: zlatko.majlinger@gmail.com, mdimitrijevic@aob.bg.ac.rs, zsimic@aob.bg.ac.rs

Spectral lines of singly charged cobalt ion (Co II) are present in stellar spectra. Such lines for example, have been observed in Hg-Mn stars (see e.g. Bolcal and Didelon, 1987), where Stark broadening is the principal pressure broadening mechanism. For Co II lines, needed for cobalt abundance determination, radiative transfer calculation and for analysis and synthesis of stellar apectra, experimental or theoretical data do not exist in literature. We started an extensive analysis of Stark broadening of spectral lines within around 50 Co II multiplets, in order to provide the needed theoretical data for Stark broadening parameters. For calculations, the modified semiempirical method (Dimitrijević and Konjević, 1980) has been used. Here, as an example of obtained results we present Stark widths for five transitions. The obtained results have been also used to demonstrate the importance of Stark broadening mechanism in white dwarfs.

References

Bolcal, C., Didelon, P.: 1987, *Elemental Abundance Analyses*, Institut d'Astronomie de l'Universite de Lausanne, Chavannes-des-Bois, Switzerland, p. 152.

Dimitrijević, M. S., Konjević, N.: 1980, *J. Quant. Spectrosc. Radiat. Transfer*, **24**, 454.

X Serbian-Bulgarian Astronomical Conference (X SBAC)
May 30 - June 3, 2016, Belgrade, Serbia
Book of Abstracts, Eds. M. S. Dimitrijević and M. K. Tsvetkov
Astronomical Observatory, Belgrade, 2016

Poster paper

PHOTOMETRIC AND SPECTRAL INVESTIGATION OF THE STAR TU UMi

D. Marchev, D. Kjurkchieva, S. Ibryamov and T. Atanasova

Department of Physics, University of Shumen,
115 Universitetska Str., Shumen, Bulgaria
E-mail: d.kyurkchieva@shu-bg.net

The paper presents BVRI photometry and high-resolution spectral observations around the H□ line of the variable star TU UMi. They were obtained correspondingly by observations with the 60-cm and 2-m telescope of NAO Rozhen in 2012-2013. The simultaneous light curve solution and radial velocity curve allowed us to obtain the global parameters of the target. The obtained results were compared with the previous ones.

Poster paper

SPECTROSCOPICAL INVESTIGATION OF GRAVITATIONALLY LENSED QUASARS

Sladjana Marčeta Mandić, Luka Č. Popović, Edi Bon and Predrag Jovanović

Astronomical Observatory, Volgina 7, 11060 Belgrade, Serbia
E-mail: sladjana.mandic@gmail.com

Here we investigate differences in redshift between images of gravitationally lensed quasars in order to study their possible correspondence to time-delays. For that purpose we selected a sample of multiply-imaged quasars and analyzed spectra of their images. Here we present preliminary results of our investigation, which show that in some cases there are slight differences in redshifts between different images of the same lensed objects.

X Serbian-Bulgarian Astronomical Conference (X SBAC)
May 30 - June 3, 2016, Belgrade, Serbia
Book of Abstracts, Eds. M. S. Dimitrijević and M. K. Tsvetkov
Astronomical Observatory, Belgrade, 2016

Poster paper

INVERSE BREMSSTRAHLUNG CHARACTERISTICS IN DWARF ATMOSPHERES: THE ABSORPTION COEFFICIENTS AND GAUNT FACTORS

Anatolij A. Mihajlov[1], Vladimir A. Srećković[1], Nenad M. Sakan[1] and Milan S. Dimitrijević[2,3,4]

[1]*University of Belgrade, Institute of Physics, P. O. Box 57, 11001, Serbia*
[2]*Astronomical Observatory, Volgina 7, 11160 Belgrade 74, Serbia*
[3]*IHIS-Technoexperts, Bežanijska 23, 11080 Zemun, Serbia*
[4]*Observatoire de Paris, 92195 Meudon Cedex, France*
E-mail: vlada@ipb.ac.rs

Here we determine the electron-ion inverse "Bremsstrahlung" characteristics for the case of the white dwarf atmospheres where such plasma characteristics as plasma density and temperature change in wide region. It is presented that determination of these characteristics i.e. the absorption coefficients and Gaunt factors can be successfully performed in the whole diapason of electron densities and temperatures which is relevant for the corresponding atmospheres.

The used quantum mechanical method of the calculation of the corresponding spectral absorption coefficient and Gaunt factor is described and discussed in details in the papers of Mihajlov *et al.* (2011, 2015).

The results are obtained for the DB White dwarf models (Koester 2015 private communication) in the wavelength region 100 nm $< \lambda <$ 3000 nm and presented in tabulated form. Also, these results can be of interest and use in investigation of different stellar and laboratory plasmas.

References

Mihajlov, A. A., Srećković, V. A., Sakan, N. M.: 2015, *Journal of Astrophysics and Astronomy,* **36**, 3.

Mihajlov, A. A., Sakan, N. M, Srećković, V. A.: 2011, *Balt. Astron.,* **20**, 604.

X Serbian-Bulgarian Astronomical Conference (X SBAC)
May 30 - June 3, 2016, Belgrade, Serbia
Book of Abstracts, Eds. M. S. Dimitrijević and M. K. Tsvetkov
Astronomical Observatory, Belgrade, 2016

Poster paper

RADIO EVOLUTION OF SUPERNOVA REMNANTS INCLUDING NON-LINEAR PARTICLE ACCELERATION

Marko Pavlović, Dejan Urošević, Bojan Arbutina, Aleksandra Dobardžić, Milica Vučetić, Vladimir Zeković and Dušan Onić

Studentski trg 16, 11000 Belgrade, Serbia
E-mail: marko@matf.bg.ac.rs

Supernova remnants (SNRs) are believed to accelerate particles up to high energies through the mechanism of diffusive shock acceleration (DSA). Radio emission from cosmic ray (CR) electrons supports this picture.

We use two-dimensional hydrodynamic simulations of SNRs coupled with particle acceleration and magnetic field amplification at non-relativistic shocks to derive the total radio emission. We take into account the dynamical reaction of the accelerated particles on the shock wave. We coupled a simple Blasi's semi-analytical model that deals with these non-linear effects in a quantitative way and changes hydrodynamics by means of an effective adiabatic index. Bell's cosmic-ray non-resonant streaming instability is considered to be responsible for the amplification of precursor magnetic field.

We obtained the radio synchrotron surface brightness increasing with time in the free expansion phase, achieving its peak value at the beginning of the Sedov phase and then decreasing during later phases. The dependence of the radio surface brightness on the diameter (time) has been calculated for different values of the interstellar medium density, supernova explosion energy, injection efficiency and different initial ejecta clumping, covering the region of the existing experimental points.

X Serbian-Bulgarian Astronomical Conference (X SBAC)
May 30 - June 3, 2016, Belgrade, Serbia
Book of Abstracts, Eds. M. S. Dimitrijević and M. K. Tsvetkov
Astronomical Observatory, Belgrade, 2016

Poster paper

DETERMINATION OF ORBITS OF VISUAL BINARY AND LINEAR ELEMENTS OF DOUBLE STARS

Rade Pavlović[1], Zorica Cvetković[1], Svetlana Boeva[2] and Goran Damljanović[1]

[1]*Astronomical Observatory, Volgina 7, 11060 Belgrade,Serbia*
[2]*Institute of Astronomy and NAO Rozhen, Bulgarian Academy of Sciences, 72 Tsarigradsko Chausse Blvd., 1784 Sofia, Bulgaria*
E-mail: rpavlovic@aob.rs

Here we provide a brief overview of methods for determination of the orbital elements for visual binary stars. Koval'skij proposed as early as 1872 an analytical method. To obtain acceptable orbital elements by this method the observational data should cover the entire orbit within at least one period. Failing this, it often happens that real elliptic solution cannot be obtained. In order to enable this method to be applied even in cases where the observations cover a shorter arc of the orbit or there are gaps Olevi´c (Olević, Cvetković 2004) introduced an improvement by using supplementary, fictive, observations.

We also provide a short description of the method for determination of linear elements for double stars which observations show a linear trend (Cvetković et al. 2012, Pavlović et al. 2013). We apply existing criteria for establishing the nature of this system, i.e. to answer to question if they are or not gravitationally bound.

We apply these methods onto several double stars which were observed at both, the Rozhen observatory and the Astronomical station at Vidojevica. Finally, we analyze the obtained results.

References

Olević, D., Cvetković, Z.: 2004, *Astronomy and Astrophysics*, **415**, 259. Pavlović, R., Cvetković, Z., Boeva S., Vince O., Stojanović, M.: 2013, *Astronomical Journal*, **146**, 52

Cvetković, Z., Pavlović, R., Boeva S., Damljanović, G.: 2012, *Bulgarian Astronomical Journal*, **18(1)**, 56.

X Serbian-Bulgarian Astronomical Conference (X SBAC)
May 30 - June 3, 2016, Belgrade, Serbia
Book of Abstracts, Eds. M. S. Dimitrijević and M. K. Tsvetkov
Astronomical Observatory, Belgrade, 2016

Poster paper

WHERE DOES DUST MATTER EXIST IN THE SYSTEM OF PLUTO–CHARON?

N. I. Perov[1] and A. A. Erokhin[2]

[1]*State Autonomous Organization of Culture "Centre named after V.V. Tereshkova", ul. Chaikovskogo, 3, Yaroslavl-150000. Russian Federation*
[2]*Yaroslavl's State Pedagogical University named after K.D. Ushinskii, ul. Respublikanskaya, 108, Yaroslavl-150000. Russian Federation*
E-mail: perov@yarplaneta.ru

Below, in accordance with work (Perov 2015), we state regions of gas and dust motion in the Pluto and Charon system. Using the model of the restricted circle planar 3 body problem "Pluto–Charon–particle of negligible mass", data (http://nssdc.gsfc.nasa.gov/planetary/planets/plutopage.html), known for the considered bodies, we with help of numerical experiments find initial conditions x_{30} (y_{30}=0, initial velocity of m_3 equals 0) for m_3 that it makes several dozen revolutions along "closed" thin curves (Fig. 1 a, b).

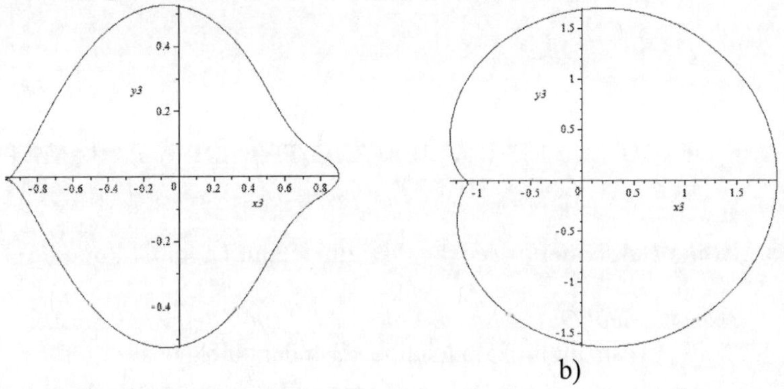

a) b)

Fig. 1. Pluto (m_1)-Charon (m_2)-particle (m_3). x_{30}=-x_2-epsilon. m_1/m_2 = 8.194968553459. x_2=0.89124487; a) ε=0.08964. 0<t<8 units of time. 100000 points; b) ε=0.19559 . 0<t<15 units of time. 10000 points. Units of length=r_{12}.

References

Perov, N. I.: 2015, *46th Lunar and Planetary Science Conference*. Abstr. No. 1021.
http://nssdc.gsfc.nasa.gov/planetary/planets/plutopage.html

X Serbian-Bulgarian Astronomical Conference (X SBAC)
May 30 - June 3, 2016, Belgrade, Serbia
Book of Abstracts, Eds. M. S. Dimitrijević and M. K. Tsvetkov
Astronomical Observatory, Belgrade, 2016

Poster paper

DETECTING SUBPARSEC SUPER-MASSIVE BINARY - LONG TERM MONITORING PERSPECTIVE

Luka Č. Popović[1], Saša Simić[3] and Nemanja Rakić[2]

[1]*Astronomical Observatory Belgrade, Volgina 7, 11060 Belgrade, Serbia*
[2]*Faculty of Science, University of Banjaluka, Mladena Stojanovića 2, 78000 Banjaluka, Republic of Srpska, Bosnia and Herzegovina*
[3]*Faculty of Sciences, University of Kragujevac, Radoja Domanovica 12, 34000 Kragujevac, Serbia*
E-mail: lpopovic@aob.rs

Here we modeled the expected spectral variability of an AGN where the supermassive binary black holes are present in the center. We consider the eclipsing effects and disc temperature variation due to the binary component interactions. Periodical and quasi-periodical line and continuum flux variations are expected and we explore the possibility of observations of these effects during spectral monitoring campaigns

Poster paper

SHAPE OF IRON Kα LINE AND ACCRETION DISK PARAMETERS IN TYPE 1 AGN

Miika Pursiainen, Predrag Jovanović and Luka Č. Popović

Astronomical Observatory, Volgina 7, 11060 Belgrade, Serbia
E-mail: miika.pursiainen@student.uibk.ac.at

To investigate the shapes of the iron Kα line detected in the X-ray spectra of Type 1 AGNs and their relation with the disk parameters, we model relativistic accretion disks around supermassive black holes. We generated several thousand of models and compared them with the *XMM-Newton* observations of the iron Kα lines in Seyfert 1 galaxies. The results show that the parameters of the disk have an important influence on the iron Kα line profiles and we discussed the fact that the broad iron Kα line is observed in the less than 50% of Type 1 AGNs.

X Serbian-Bulgarian Astronomical Conference (X SBAC)
May 30 - June 3, 2016, Belgrade, Serbia
Book of Abstracts, Eds. M. S. Dimitrijević and M. K. Tsvetkov
Astronomical Observatory, Belgrade, 2016

Poster paper

THE MAYAN CALENDAR – FACTS AND DOUBTS

Kristina Racković Babić, Dušan Marčeta and Stevo Šegan

University of Belgrade,Faculty of Mathematics, Studentski trg 16, Belgrade, Republic of Serbia
E-mail: rackovick@yahoo.com

The Mayan calendar consists of several cycles or counts of the different lengths. Practically, it is real to assume that Maya have had several calendar systems, more or less independent of each other. Common thing for all of them is unknown correspondence among their primary dates (zero point of calendar) and so called era of the European calendars. In this paper we try to explain some details of Mayan calendar practice, on the base of the survived codices – Dresden's, Madrid's and Paris, and, in addition, we will try to find conditions and rules of their using derived as a result of ancient astronomical facts.

X Serbian-Bulgarian Astronomical Conference (X SBAC)
May 30 - June 3, 2016, Belgrade, Serbia
Book of Abstracts, Eds. M. S. Dimitrijević and M. K. Tsvetkov
Astronomical Observatory, Belgrade, 2016

Poster paper

PROCEDURE FOR DETECTION OF CHARACTERISTIC RADIO SIGNAL VARIATIONS INDUCED BY SOLAR X-RAY FLARES

Dušan Raičević[1], Jovan Bajčetić[1], Vladimir M. Čadež[2] and Aleksandra Nina[3]

[1]*Department of Telecommunications and Information Science, University of Defence, Military Academy, Belgrade, Serbia*
[2]*Astronomical Observatory of Belgrade, Belgrade, Serbia*
[3]*Institute of Physics, University of Belgrade, Belgrade, Serbia*
E-mail: ricek9@gmail.com

To make a general study of characteristic variations of radio signals induced by a given type of perturber it is necessary to do a relevant statistical analysis which requires automatization of signal processing (Nina *et al.* 2015).

This paper shows a procedure for determination of characteristic parameters of signal perturbation induced by solar X-ray flares as one of the most important sources of perturbation of the low ionosphere and, consequently, the electromagnetic wave propagation in this medium (Bajčetić *et al.* in press). These variations are result from the electron density increase during such an event. Here we apply the developed procedure on amplitude values of the signal emitted by the 23.4 kHz DHO radio transmitter in Germany and received by the AWESOME receiver in the Institute of Physics in Belgrade, Serbia. The outputs of the procedure are times of the start, maximum and end of the amplitude perturbation, corresponding amplitudes, as well as their differences.

References

Bajčetić, J., Nina, A., Čadež, V. M., Todorović, B. M.: *Thermal Science,* in press, doi:10.2298/TSCI141223084B.

Nina, A., Simić, S., Srećković, V. A., Popović, L. Č.: 2015, *Geophysical Research Letters,* **42** (19), 8250.

X Serbian-Bulgarian Astronomical Conference (X SBAC)
May 30 - June 3, 2016, Belgrade, Serbia
Book of Abstracts, Eds. M. S. Dimitrijević and M. K. Tsvetkov
Astronomical Observatory, Belgrade, 2016

Poster paper

OPTICAL SPECTRAL VARIABILITY OF QUASAR E1821+643

A. I. Shapovalova[1], L. Č. Popović[2,3], D. Ilić[3], V. Chavushyan[4], A. N. Burenkov[1], W. Kollatschny[5] and A. Kovačević[3]

[1]*Special Astrophysical Observatory of the Russian AS, Nizhnij Arkhyz, Karachaevo-Cherkesia 369167, Russia*
[2]*Astronomical Observatory, Volgina 7, 11160 Belgrade 74, Serbia*
[3]*Department of Astronomy, Faculty of Mathematics, University of Belgrade, Studentski trg 16, 11000 Belgrade, Serbia*
[4]*Instituto Nacional de Astrofisica, Óptica y Electrónica, Apartado Postal 51-216, 72000 Puebla, Puebla, México*
[5]*Institut für Astrophysik, Georg-August-Universität Göttingen, Germany*
E-mail: dilic@matf.bg.ac.rs

The quasar E1821+643 is an interesting object since it is suggested that it could host a binary black hole in its center. Therefore we have performed a long term (1990–2014) spectrophotometric monitoring of this object, that has been done for the first time. Here we report our major findings of the variability of the continuum and line fluxes.

We found periodical variations in the photometric flux with periods of 1200, 1850, and 4000 days, and 4500-day periodicity in the spectroscopic variations. While the continuum and line fluxes are varying during the monitoring period, the line profiles have not significantly changed, but have always a red asymmetry and broad line peak redshifted around +1000 km s^{-1}.

The obtained results are discussed in the frame of the binary black hole hypothesis (Shapovalova *et al.* 2016).

References

Shapovalova, A. I., Popović, L. Č., Chavushyan, V., *et al.*: 2016, *ApJS,* **222**, 25.

X Serbian-Bulgarian Astronomical Conference (X SBAC)
May 30 - June 3, 2016, Belgrade, Serbia
Book of Abstracts, Eds. M. S. Dimitrijević and M. K. Tsvetkov
Astronomical Observatory, Belgrade, 2016

Poster paper

CONTRIBUTION TO SED OF AGNs INDUCED BY POSSIBLE DENSITY PERTURBATIONS IN COMPLEX GEOMETRY OF BINARY SYSTEMS

Saša Simić[1] and Edi Bon[2]

[1]*Faculty of Science, Department of Physics, University of Kragujevac, Serbia*
[2]*Astronomical Observatory Belgrade, Volgina 7, 11060 Belgrade, Serbia*
E-mail: ssimic@kg.ac.rs

Here we test effects of emission from specific configurations of binary black hole systems, as a source of continuum flux of such objects. We consider that orbital motion of BHs in dense environment can induce density perturbation in form of a spirals that are expected to form in the inner part of circum binary disk to each BH due to gap openings. We compute the output of the binary BH emission considering the complex geometry which include the mini accretion disks around each BH accompanied with extended spiral arms.

X Serbian-Bulgarian Astronomical Conference (X SBAC)
May 30 - June 3, 2016, Belgrade, Serbia
Book of Abstracts, Eds. M. S. Dimitrijević and M. K. Tsvetkov
Astronomical Observatory, Belgrade, 2016

Poster paper

HF ELECTRIC PROPERTIES OF THE ASTROPHYSICAL PLASMAS

Vladimir A. Srećković[1], Anatolij A. Mihajlov[1], Nenad M. Sakan[1], Ljubinko M. Ignjatović[1], Milan S. Dimitrijević[2,3,4], Darko Jevremović[2] and Veljko Vujčić[2]

[1]*University of Belgrade, Institute of Physics, P. O. Box 57, 11001, Serbia*
[2]*Astronomical Observatory, Volgina 7, 11160 Belgrade 74, Serbia*
[3]*IHIS-Technoexperts, Bežanijska 23, 11080 Zemun, Serbia*
[4]*Observatoire de Paris, 92195 Meudon Cedex, France*
E-mail: vlada@ipb.ac.rs

Here we determine the HF characteristics of astrophysical plasmas on the basis of numerically calculated values for the dense plasma conductivity in an external HF electric field. The examined range of plasma frequencies covers the IR, visible and UV regions and consider electronic number density and temperature important for different stellar models.

The results presented here are important for the investigation of atmosphere plasmas of astrophysical objects like white dwarfs with different atmospheric compositions (DA, DC etc.), and for investigation of some other stars (M-type red dwarfs, Sun etc.) as well as for laboratory plasma research (Srećković *et al.* 2010).

References

Srećković, V. A., Ignjatović, Lj. M., Mihajlov, A. A., Dimitrijević, M. S: 2010, *MNRAS,* **406**, 590.

X Serbian-Bulgarian Astronomical Conference (X SBAC)
May 30 - June 3, 2016, Belgrade, Serbia
Book of Abstracts, Eds. M. S. Dimitrijević and M. K. Tsvetkov
Astronomical Observatory, Belgrade, 2016

Poster paper

DISTURBANCES IN THE D-REGION INDUCED BY LARGE SOLAR FLARES

Vladimir A. Srećković[1], Desanka M. Šulić[2] and Anatolij A. Mihajlov[1]

[1]*University of Belgrade, Institute of Physics, P. O. Box 57, 11001, Serbia*
[2]*University Union - Nikola Tesla, 11000, Belgrade, Serbia*
E-mail: vlada@ipb.ac.rs

Observed amplitude and phase on Very Low Frequency (VLF) radio signals can be used to study variability, morphology and other phenomena occurring in the D-region of the ionosphere. Great variations in amplitude and phase occur near sunrise and sunset, while more steady levels of amplitude are observed during daytime and nighttime conditions. Among many natural phenomena effects of solar flares can be observed by perturbations of amplitude and phase on VLF radio signals (Šulić *et al.* 2014, 2016). On the basis of measured intensity of solar X-ray by GOES satellites data, the X17.2 ($I_X = 1.72 \cdot 10^{-3}$ Wm^{-2} X-ray flux in the band 01-0.8 nm) class solar flare is ranking on the third place on the list of the most powerful solar flares recorded since 1976. Beside this solar flare for studying the influence of large solar X-ray flare on the enhancement of the D-region electron density the X5.4 ($I_X = 5.4 \cdot 10^{-4}$ Wm^{-2}) class solar flare is selected. Both of these flares were recorded in October 2003. Simultaneous amplitude and phase on GQD/22.10 kHz radio signal measurements made over short (~ 2000 km) mainly land path are used to determine the enhancements of electron density in the D-region induced by large solar X-ray flares.

References

Šulić, D. M., Srećković, V. A.: 2014, *Serb. Astr. Jour.* **188**, 45.
Šulić, D. M., Srećković, V. A., Mihajlov, A. A: 2016, *Adv. Space Res.* **57**, 85.

X Serbian-Bulgarian Astronomical Conference (X SBAC)
May 30 - June 3, 2016, Belgrade, Serbia
Book of Abstracts, Eds. M. S. Dimitrijević and M. K. Tsvetkov
Astronomical Observatory, Belgrade, 2016

Poster paper

OPTICAL MONITORING OF QSOS IN THE FRAMEWORK OF THE GAIA SPACE MISSION

F. Taris[1], G. Damljanović[2], A. Andrei[3,4,1], J. Souchay[1] and O. Becker[1]

[1]*Observatoire de Paris - SYRTE, PSL Research University, CNRS/UMR8630, Sorbonne Universités, Université Pierre et Marie Curie, LNE, 61 avenue de l'Observatoire, 75014, Paris, France*
[2]*Astronomical Observatory, Volgina 7, 11060 Belgrade, Serbia*
[3]*Observatorio Nacional/MCT, Rio de Janeiro, Brasil*
[4]*Observatorio do Valongo, UFRJ, Brasil*
E-mail: francois.taris@obspm.fr

This poster presents the set of optical telescopes used to observe the targets chosen for the link between the ICRF and the future GCRF. We will focus on results obtained with ASV 60 cm instrument (Serbia, photometry) and 2m Rozhen telescope (Bulgaria, morphology).

Moreover we present the results of the Lomb-Scargle and CLEAN methods applied to the data obtained with the TAROT robotic telescopes (OCA, France, and ESO, Chile). We show that the same periods are obtained by both methods with a high confidence level.

Finally we describe a project of a database currently under construction to gather information (optical images, photometry, morphology) about the quasars regularly observed in the radio domain (ICRF targets) and that are part of the LQAC catalogue. The images are coming from the set of optical telescopes dedicated to these observations, from the data mining of observatory data bases (CFHT, SUBARU, ESO) or from the DSS/SDSS surveys. This database will be publically available through the ICRS-PC website.

X Serbian-Bulgarian Astronomical Conference (X SBAC)
May 30 - June 3, 2016, Belgrade, Serbia
Book of Abstracts, Eds. M. S. Dimitrijević and M. K. Tsvetkov
Astronomical Observatory, Belgrade, 2016

Poster paper

ASTRONOMICAL EDUCATION IN THE CULTURAL AND EDUCATIONAL CENTRE NAMED AFTER V. V. TERESHKOVA

Ekaterina N. Tikhomirova

State Autonomous Culture Institution of Yaroslavl region «Cultural and educational centre named after V.V.Tereshkova», Tchcaikovskogo st., 3, Yaroslavl, Russia, 150000
E-mail: en_tihomirova@mail.ru

The educational technologies, connecting with the astrospace researches in one way or another are particularly topical nowadays. Planetaria are considered as the unique centers where educational and research activities are successfully combined.

Cultural and educational centre is a modern cultural and educational complex including the planetarium, the museum of the history of cosmonautics, the astronomical observatory, the interactive classroom for astronomical study group, media café "Trans – force". The outreach and educational work is carried out with a population of different age and social groups on the basis of the Centre (Tikhomirova, Gorodenskaya 2015).

The educational process assumes using the richest material of the museum "History of Cosmonautics" of Valentina Tereshkova Cultural and educational centre and other museums. The Star hall (planetarium) of the Centre allows for any weather, at any time of day and year, at intervals of thousands of years - "travel to the future and the past" – to demonstrate the starry sky, and observed in different parts of the globe. Besides the starry sky, there are ample opportunities for studying of Solar system, planets, and comets, fireballs, meteor showers, and others. The deep sky objects can also be represented using specialized software and digital optical systems. Thus, the proper use of the grand demonstration opportunities of planetarium using traditional and innovative techniques is the guarantee not only of high interest to the presented material, but also attraction of youth to activity in space and astronomical researches. The astronomical observatory of the Centre promotes to the acquisition of practical skills for working with astronomical devices.

References

Tikhomirova, E. N., Gorodenskaya, M. N.: 2015, *LPS* **XXXXVI**, Abstract # 1390.

X Serbian-Bulgarian Astronomical Conference (X SBAC)
May 30 - June 3, 2016, Belgrade, Serbia
Book of Abstracts, Eds. M. S. Dimitrijević and M. K. Tsvetkov
Astronomical Observatory, Belgrade, 2016

Poster paper

REAL TIME VARIATIONS OF IONOSPHERIC TEC (TOTAL ELECTRON CONTENT) DURING SOLAR X-RAY FLARES

Miljana Todorović Drakul[1], Jovan Bajčetić[2], Vladimir M. Čadež[3], Luka Č. Popović[3], Dragan M. Blagojević[1] and Aleksandra Nina[4]

[1]*Department of Geodesy and Geoinformatics, Faculty of Civil Engineering, University of Belgrade, Belgrade, Serbia*
[2]*Department of Telecommunications and Information Science, University of Defence, Military Academy, Belgrade, Serbia*
[3]*Astronomical Observatory of Belgrade, Belgrade, Serbia*
[4]*Institute of Physics, University of Belgrade, Belgrade, Serbia*
E-mail: mtodorovic@grf.bg.ac.rs

The total number of electrons integrated along a tube of unit cross section or the electron columnar density is called TEC (Total Electron Content) and, in this paper, our attention is focused on its changes during the occurrence of solar X-ray flares.

This research is concentrated primarily to the D-region of the ionosphere and its contribution to changes of TEC. As examples for quantitative calculations of D-region reactions to solar X-ray flares we chose events that occurred on May 5th, 2010, February 18th, 2011, and March 24th, 2011.

The ionospheric modelling is based on experimental data obtained by the low ionosphere observations using the 23.4 kHz VLF (the very low frequency) radio signal emitted in Germany and received in Serbia (also used in Nina *et al.* 2011, 2013 and Bajčetić *et al.* in press) while GPS was used to determine the vertical TEC.

References

Bajčetić, J., Nina, A., Čadež, V. M., Todorović, B. M.: *Thermal Science,* in press, doi:10.2298/TSCI141223084B.

Nina, A., Čadež, V. M., Srećković, V., Šulić, D.: 2011, *Baltic Astronomy,* **20**, 609.

Nina, A., Čadež, V. M.: 2013, *Geophysical Research Letters,* **40** (18), 4803.

X Serbian-Bulgarian Astronomical Conference (X SBAC)
May 30 - June 3, 2016, Belgrade, Serbia
Book of Abstracts, Eds. M. S. Dimitrijević and M. K. Tsvetkov
Astronomical Observatory, Belgrade, 2016

Poster paper

INCORPORATION OF THE BELGRADE OBSERVATORY ZEISS ASTROGRAPH PLATE ARCHIVE IN THE WFPDB

Milcho Tsvetkov[1], Damyan Kalaglarsky[1], Katya Tsvetkova[1], Darko Jevremović[2], Vojislava Protić-Benišek[2], Vladimir Benišek[2] and Milan S. Dimitrijević[2]

[1]*Institute of Mathematics and Informatics, Bulgarian Academy of Sciences, Acad. G. Bonchev Str. 8, Sofia-1113, Bulgaria*
[2]*Astronomical Observatory, Volgina 7, 11060 Belgrade, Serbia*
E-mail: milcho.tsvetkov@gmail.com

Here we present the results of the incorporation n the Wide-Field Plate Database (WFPDB, Tsvetkov and Tsvetkova, 2012, www.wfpdb.org) of the historical plate archive of the 16 cm Zeiss Astrograph of the Belgrade observatory published on the web site of the Serbian Virtual Observatory (SerVO, http://servo.aob.rs/photo_plates_archive/). This work is a result of the long term cooperation between Bulgarian and Serbian astronomers last decades (Tsvetkova et al. 2010). The plate archive of the Belgrade Zeiss Astrograph contains 1939 plates digitized with the Epson V700 Perfection flatbed scanner. The plates were taken in the period 1936-1973 mainly by M. Protić and Č. Čepinac. Analysis of the incorporated plate archive is presented. The plate archive is available online in WFPDB-Sofia and will be linked to SerVO and International Virtual Observatory.

References

Tsvetkov, M., Tsvetkova, K.: 2012, in *Proceedings of the International Astronomical Union Symposium No. 285 "New Horizons in Time-Domain Astronomy"*, Eds. R. E. M. Griffin, R. J. Hanisch, R. Seaman, p. 417-419.

Tsvetkova, K., Tsvetkov, M., Dimitrijević, M. S., Protić-Benišek, V., Benišek, V., Jevremović D.: 2010, *Mem. S.A.It. Suppl.*, **15**, 192.

X Serbian-Bulgarian Astronomical Conference (X SBAC)
May 30 - June 3, 2016, Belgrade, Serbia
Book of Abstracts, Eds. M. S. Dimitrijević and M. K. Tsvetkov
Astronomical Observatory, Belgrade, 2016

Poster paper

LIGHT CURVE SOLUTIONS OF THE ECCENTRIC *KEPLER* BINARIES KIC 4281895 AND KIC 5115178 WITH TIDALLY INDUCED HUMPS

Doroteya Vasileva and Diana Kjurkchieva

Department of Physics, University of Shumen, 115 Universitetska Str., Shumen, Bulgaria
E-mail: d.vasileva@shu.bg, d.kyurkchieva@shu-bg.net

We carried out light curve solutions of two eclipsing detached binaries on eccen-tric orbits observed by *Kepler*. The orbits and stellar parameters of KIC 4281895 and KIC 5115178 were determined with a high accuracy by modeling of their photometric data. We found also tidally induced brightening (hump) around the periastron phase of our targets.

Poster paper

OBSERVATIONS OF NGC185 GALAXY – STUDY OF SUPERNOVA REMNANT IN A DWARF ELLIPTICAL GALAXY

M. M. Vučetić[1], B. Arbutina[1], M. Z. Pavlović[1], A. Ćiprijanović[1], D. Urošević[1], N. Petrov[2], D. Onić[1] and A. Trčka[1]

[1]*Department of Astronomy, Faculty of Mathematics, University of Belgrade, Studentski trg 16, 11000 Belgrade, Serbia*
[2]*National Astronomical Observatory Rozhen, Institute of Astronomy, Bulgarian Academy of Sciences, 72 Tsarigradsko Shosse Blvd, BG-1784 Sofia, Bulgaria*
E-mail: mandjelic@matf.bg.ac.rs

Due to the lack of interstellar matter, supernova remnants (SNRs) are rarely observed in elliptical galaxies. In this paper we discuss the previously known optical supernova remnant (SNR) in NGC185 galaxy, a dwarf elliptical companion of the Andromeda galaxy. We observed a central portion of NGC185, through the narrowband Hα and [SII] filters, on a 2m RCC-telescope at National astronomical observatory Rozhen, Bulgaria. Finally, we performed MHD simulation using the Pluto code and compared standard evolution models with the case of low environmental density and high pressure, in order to determine properties and parameters of the SNR and its host galaxy.

X Serbian-Bulgarian Astronomical Conference (X SBAC)
May 30 - June 3, 2016, Belgrade, Serbia
Book of Abstracts, Eds. M. S. Dimitrijević and M. K. Tsvetkov
Astronomical Observatory, Belgrade, 2016

Poster paper

OPTICAL OBSERVATIONS OF NGC1156 GALAXY IN NARROW BAND [S II] AND H*α* FILTERS

M. M. Vučetić[1], A. Trčka[1], B. Arbutina[1], A. Ćiprijanović[1], M. Z. Pavlović[1], D. Urošević[1] and N. Petrov[2]

[1]*Department of Astronomy, Faculty of Mathematics, University of Belgrade, Studentski trg 16, 11000 Belgrade, Serbia*
[2]*National Astronomical Observatory Rozhen, Institute of Astronomy, Bulgarian Academy of Sciences, 72 Tsarigradsko Shosse Blvd, BG-1784 Sofia, Bulgaria*
E-mail: mandjelic@matf.bg.ac.rs

We present observations of the Magellanic-type dwarf irregular galaxy NGC1156 in narrow band [SII] and Hα filters, carried out in November 2015 with the 2m RCC telescope at Rozhen National Astronomical Observatory in Bulgaria. Although our observations were with the intention to find new extragalactic supernova remnant candidates, we were successful only in the detection of HII regions.

Poster paper

EVOLUTION IN THE SYSTEM CORONA-DISK

Krasimira Dimitrova Yankova

Space Research and Technology Institute - BAS
Acad G. Bonchev Str., Bl. 1, 1113 Sofia, Bulgaria
E-mail: f7@space.bas.bg

This paper considers magneto-hydrodynamics of the system disk-corona. We will researching the structuring of the flow on the secondary component. Analyze the influence of the destributions on boundery with the primary component. Discusses the significance of the type of the border for exchange of energy between components.

X SERBIAN-BULGARIAN ASTRONOMICAL CONFERENCE

May 30 - June 3, 2016
Faculty of Civil Engineering, Belgrade, Serbia

PROGRAMME

30. 05. 2016, Monday

18:00-19:00 Registration

19:00 WELCOME COCKTAIL

31. 05. 2016, Tuesday

09:00-09:30 Opening Ceremony

Chairman: Milcho Tsvetkov

09:30-10:00 V. L. Afanasiev, L. Č. Popović: INVESTIGATION OF SUB-PC AGN POLARIZATION REGIONS AND GRAVITATIONAL LENSES WITH OPTICAL SPECTROPOLARIMETRIC METHOD

10:00-10:30 Ognyan Kounchev: DATA ANALYTICS FOR BIG DATA IN ASTROINFORMATICS

10:30-11:00 V. Borka Jovanović, P. Jovanović, D. Borka, S. Capozziello: R^N GRAVITY AS VIABLE ALTERNATIVE TO DARK MATTER: APPLICATION TO STELLAR DYNAMICS

11:00-11:30 Coffee Break

EXTRAGALACTIC ASTRONOMY

Chairman: Luka Č. Popović

11:30-12:00 Alla Shapovalova, Nikolaj Bochkarev, Luka Č. Popović et al.: CHARACTERISTICS OF THE LONG-TERM SPECTRAL VARIABILITY OF THE AGNs WITH BROAD LINES IN THE OPTICAL SPECTRAL BAND

12:00-12:30 Jelena Kovačević Dojčinović, Luka Č. Popović: SPECTRAL CHARACTERISTICS OF THE AGNs TYPE 1 IN THE UV/OPTICAL SPECTRAL BAND

12:30-12:45 Rosa Victoria Munoz Dimitrova, Rumen Bachev: OPTICAL VARIABILITY OF DIFFERENT TYPES OF AGNs OBSERVED FROM ROZHEN OBSERVATORY

12:45-13:00 Djordje V. Savić, Rene W. Goosmann, Martin Gaskell, Frederic Marin, Luka Č. Popović: POLARIZATION IN AGN BROAD EMISSION LINES – THE CENTRAL SOURCE ANIZOTROPY AND GAS KINEMATICS

13:00-13:15 Andjelka Kovačević, Luka Č. Popović, Alla I. Shapovalova, Dragana Ilić: GAUSSIAN PROCESSES FOR LONG-TERM OBSERVED CONTINUUM AND BROAD LINE LIGHT CURVES OF E1821+643

13:15-13:30 Nataša Bon, Edi Bon, Paola Marziani: NEW METHOD TO STUDY AGN SPECTRAL VARIABILITY

13:30-13:45 Edi Bon, Paola Marziani, Nataša Bon: THE TWO COMPONENT MODEL OF THE BLR OPTICAL EMISSION IN THE NGC 5548

13:45-14:00 Marijana Smailagić, Edi Bon: SIMPLIFIED MODEL OF LINE PROFILE VARIABILITY FROM ECCENTRIC ORBITS OF SUPERMASSIVE BINARY BLACK HOLE SYSTEMS

14:00-15:30 Lunch Break

SUN

Chairman: Petko Nedialkov

15:30-16:00 Vladimir N. Obridko, I. V. Dmitrieva: RELATIONSHIP BETWEEN SOLAR MAGNETIC FIELDS OF DIFFERENT SPATIAL SCALES AND SOME CONTROVERSIAL PROBLEMS OF THE DYNAMO THEORY

16:00-16:30 M. Dechev, Ts. Tsvetkov, K. Koleva, P. Duchlev: A SOLAR PROMINENCE ERUPTION FROM THREE SPACECRAFT POINTS OF VIEW

16:30-17:00 Aleksandra Nina: IONOSPHERIC PERTURBATIONS INDUCED BY SOLAR X-RAY FLARES

17:00-17:15 Ts. Tsvetkov, K. Koleva, P. Duchlev, M. Dechev: MULTI-WAVELENGTH OBSERVATIONS OF A TWO-RIBBON SOLAR FLARE CAUSED BY FILAMENT ERUPTION

17:15-17:30 D. M. Šulić, V. A. Srećković, A. A. Mihajlov: LARGE SOLAR FLARES AND THEIR IMPACT ON THE IONOSPHERE

17:30-18:00 Coffee Break

ASTRONOMY IN CULTURE AND SOCIETY

Chair: Nadežda Pejović

18:00-18:15 P. Nedialkov, G. Andrejević: COMETS – FIERY SWORD CONNECTION: A KRUŠEDOL CLUE

18:15-18:30 Ивар Шмелд: РАДИОАСТРОНОМИЯ В ЛАТВИИ – ИСТОРИЯ И ПЕРСПЕКТИВЫ РАЗВИТИЯ

18:30-18:45 Jovan Aleksić: ASTRONOMICAL SOCIETY "RUDJER BOSKOVIC"

18:45-19:00 Vladimir Đošović, Bojana Tajhmeister, Kristina Pantelić: ASTRONOMICAL CONCERNS AND THEIR SOLVING IN CONNECTION WITH THE ARCHEO-ASTRONOMICAL LOCALITY LEPENSKI VIR

01. 06. 2016, Wednesday

09:00-12:0 Meeting of Euroasian Astronomical Society (EAAS)
Meeting of EAAS Council Board

14:30 SOCIAL PROGRAM

02. 06. 2016, Thursday

ASTROINFORMATICS

Chairman: Ognyan Kounchev

09:00-9:30 K. Tsvetkova, M. Tsvetkov, N. Kirov, D. Kalaglarsky, H. Edelmann, U. Heber: ON SOME BAMBERG WIDE-FIELD PLATE CATALOGUES RECENTLY INCORPORATED INTO WFPDB

9:30-10:00 N. N. Samus, S. V. Antipin, D. M. Kolesnikova, K. V. Sokolovsky, M. K. Tsvetkov: SCANNING MOSCOW PLATE COLLECTION: PROGRAM, ELECTRONIC CATALOGS, AND NEW VARIABLE STARS IN THE FIELD OF 104 Her

10:00-10:30 Lasko Laskov: METHODS FOR DOCUMENT IMAGE DE-WARPING

10:30-10:45 N. Kirov, M. Tsvetkov, K. Tsvetkova: SOFTWARE TOOL FOR TIME CONVERSION IN THE WFPDB

10:45-11:00 Vladimir Benishek, Vojislava Protitch Benishek: SERBIAN VIRTUAL OBSERVATORY: BELGRADE ASTROPHOTOGRAPHIC PLATE ARCHIVE OF SOLAR OBSERVATIONS

11:00-11:30 Coffee Break

ASTROMETRY

Chairman: Jan Vondrák

11:30-12:00 Werner W. Zeilinger: GAIA SCIENCE ALERTS FOLLOW-UP OBSERVATIONS

12:00-12:30 Goran Damljanović, Oliver Vince, Miljana D. Jovanović, Zorica Cvetković, Rade Pavlović, Svetlana Boeva, Georgi Latev, Milan Stojanović, Gabrijela Marković: SERBIAN-BULGARIAN MINI-NETWORK TELESCOPES AND GAIA-FUN-TO

12:30-13:00 S. Boeva, G. Damljanović, B. Petrov, B. Spassov, M. Sekulić, G. Latev: SERBIAN – BULGARIAN MINI – NETWORK TELESCOPES: FIRST SIMULTANEOUS OBSERVATIONS OF VARIABLE OBJECTS

13:00-13:15 Grigor Nikolov: PRECISE ASTROMETRY IN STAR CLUSTERS FROM CENTURY LONG OBSERVATIONS

13:15-15:00 Lunch Break

LABORATORY ASTROPHYSICS, SPACE PLASMAS AND FIELDS

Chairman: Nikolai N. Samus

15:00-15:30 Vladimir A. Srećković, Anatolij A. Mihajlov, Ljubinko, M. Ignjatović, Milan S. Dimitrijević: ATOM-RYDBERG ATOM PROCESSES IN THE STELLAR ATMOSPHERES: DWARF ATMOSPHERES, QUIET SUN AND SUNSPOTS

15:30-15:45 R. Beuc, G. Peach, M. Movre, B. Horvatić: LITHIUM, SODIUM AND POTASSIUM RESONANCE LINES PRESSURE BROADENED BY HELIUM ATOMS

15:45- 16:00 Sava Donkov, Ivan Stefanov: DENSITY DISTRIBUTION FUNCTION IN A SELF-GRAVITATING INCOMPRESSIBLE TURBULENT FLUID

16:00-16:15 E. A. Mikhailov, V. V. Pushkarev: FLUCTUATING GOVERNING PARAMETERS IN GALAXY DYNAMO

16:15-16:30 M. M. Martinović, A. Zaslavsky, M. Maksimović, S. Šegan: VENUS IONOSPHERE ELECTRON PROPERTIES – CASSINI QUASI-THERMAL NOISE MEASUREMENTS

16:30-17:00 Coffee Break

EARTH

Chairman: Žarko Mijajlović

17:00-17: 30 Jan Vondrák, Cyril Ron, Yavor Chapanov: NEW DETERMINATION OF PERIOD AND QUALITY FACTOR OF CHANDLER WOBBLE, CONSIDERING GEOPHYSICAL EXCITATIONS

17:30-17:45 Yavor Chapanov, Ivan Georgiev: EARTH ORIENTATION PARAMETERS AND GRAVITY VARIATIONS DETERMINED FROM LAGEOS 1 AND LAGEOS 2 DATA FOR THE PERIOD 1984 – 2011

17:45-18:00 Vladimir M. Čadež, Aleksandra Nina: MHD WAVES IN THE LOWER TERRESTRIAL IONOSPHERE

20:00 CONFERENCE DINNER

03. 06. 2016, Friday

STARS

Chair: Diana Kjurkchieva

09:30-10:00 Gantcho Gantchev, Petko Nedialkov, Valentin Ivanov, Evgeni Ovcharov, Antoniya Valcheva and Milen Minev: PHOTOMETRIC VARIABILITY OF LUMINOUS BLUE VARIABLES IN M33 ON SHORT TIMESCALES

10:00-10:30 Igor F. Malov: NEWS OF PULSAR ASTRONOMY

10:30-11:00 E. A. Karitskaya, N. G. Bochkarev, V. P. Goranskij, N. V. Metlova: ON LONG-TERM CYG X-1=V1357 CYG VARIABILITY

11:00-11:15 Daniela Boneva: EMISSION PROPERTIES OF WHITE DWARF'S ACCRETING BINARIES BY THE POLARIZATION METHODS

11:15-11:30 G. Latev, V. Popov, S. Boeva, P. Nikolov, B. Spassov: WZ SGE TYPE CATACLYSMIC VARIABLE ASASSN-14CL: SUPERHUMPS AND FLICKERING IN 2014 SUPEROUTBURST

11:30-12:00 Coffee Break

COSMOLOGY AND ISM

Chairman: Robert Beuc

12:00-12:15 D. Borka, P. Jovanović, V. Borka Jovanović, S. Capozziello: CONSTRAINTS ON SANDERS GRAVITY FROM PRECESSION OF ORBITS OF S2-LIKE STARS: A CASE OF A BULK DISTRIBUTION OF MASS

12:15-12:30 Ivan Dimitrijević: NONSINGULAR BIG BANG IN NONLOCAL MODIFIED GRAVITY

12:30-12:45 Žarko Mijajlović, Nadežda Pejović: ASYMPTOTIC SOLUTION FOR SPATIALLY OPEN UNIVERSE WITH MATTER-DOMINATED EVOLUTION

12:45-13:00 O. Stanchev, T. Veltchev, S. Donkov: DENSITY SCALING RELATION IN ORION A: EFFECTS OF REGION SELECTION

Chair: Katya Tsvetkova

13:00-15:00 POSTER PRESENTATION 2-3 minutes per poster

15:00 Closing Ceremony

POSTERS

1. Bojan Arbutina: SELECTED SERBIAN ASTRONOMERS IN THE WEB OF SCIENCE IN THE PERIOD 1977-2014

2. K. N. Arefieff, N. N. Bezuglov, M. S. Dimitrijević, A. N. Klyucharev, A. A. Mihajlov, V. A. Srećković: ON THE ANOMALOUS LOW SPONTANEOUS EMISSION RATES FOR p-SERIES OF SODIUM DUE TO THE EFFECT OF NATURAL FÖRSTER RESONANCE

3. Teodora Atanasova, Diana Kjurkchieva: LIGHT CURVE SOLUTIONS OF FOUR DETACHED ECLIPSING *KEPLER* BINARIES

4. Jovan Bajčetić, Aleksandra Nina: INFLUENCE OF SOLAR X-RAY FLARES ON RADIO SIGNAL PROPAGATION IN THE LOW IONOSPHERE

5. Igor I. Baluk, Ivan M. Sergei: 20 NEW VARIABLE STARS DETECTED BY CRTS

6. Borislav Stanishev Borisov: A SIMILARITY BETWEEN THE SOLAR SYSTEM AND THE KEPLER'S THREE-PLANETARY EXTRASOLAR SYSTEMS

7. Yavor Chapanov, Cyril Ron, Jan Vondrak: SOLAR INFLUENCE ON SUBDECADAL VARIATIONS OF EARTH ROTATION, MSL AND CLIMATE, DERIVED BY HARMONICS OF HALE, JOSE, DE VRIES AND SUESS CYCLES

8. Goran Damljanović, Oliver Vince, Rumen Bachev, Werner W. Zeilinger: WEBT PROJECT AND OBSERVATIONS AT THE 60 cm ASV TELESCOPE

9. Milan S. Dimitrijević, Nenad Milovanović, Zoran Simić, Sylvie Sahal-Bréchot: STARK BROADENING PARAMETERS OF S II SPECTRAL LINES FOR STELLAR PLASMA CONDITIONS

10. Milan S. Dimitrijević, Zoran Simić, Sylvie Sahal-Bréchot: ON THE STARK BROADENING OF Si III SPECTRAL LINES IN STELLAR PLASMA

11. Milan S. Dimitrijević, Zoran Simić, Andjelka Kovačević, Sylvie Sahal-Bréchot: ON THE INFLUENCE OF STARK BROADENING ON Xe VIII SPECTRAL LINES IN DO WHITE DWARFS

12. Ana Djulaković, Aleksandra Nina, Saša Simić, Vladimir A. Srećković, Luka Č. Popović: SHORT-TERM PERTURBATIONS IN HIGH AND MIDDLE LATITUDE LOW IONOSPHERE UNDER EUROPE INDUCED BY GRB_S

13. Sunay Ibryamov, Evgeni Semkov, Stoyanka Peneva: MULTICOLOUR PHOTOMETRIC STUDY OF T TAURI VARIABLES

14. Sunay Ibryamov, Evgeni Semkov, Stoyanka Peneva: LONG-TERM UBVRI PHOTOMETRY OF THE PMS STAR V350 CEPHEI

15. Natalija Janc, Vojislava Protić Benišek, Vladimir Benišek, Milivoj B. Gavrilov, Luka Č. Popović, Slobodan B. Marković: ACADEMICIANS MILUTIN MILANKOVIĆ AND VOJISLAV MIŠKOVIĆ – CORRESPONDENCE

16. Darko Jevremović, Veljko Vujčić, Anatolij A. Mihajlov, Vladimir A. Srećković, Ljubinko M. Ignjatović, Milan S. Dimitrijević: MOL-D DATABASE FOR MOLECULAR COLLISIONAL AND RADIATIVE PROCESSES

17. Diana Kjurkchieva, V. Popov, Doroteya Vasileva, N. Petrov: OBSERVATIONS AND LIGHT CURVE SOLUTIONS OF THE ECLIPSING BINARIES USNO-B1.0 1395-0370184 AND USNO-B1.0 1395-0370731

18. Zlatko Majlinger, Milan S. Dimitrijević, Zoran Simić: ON THE STARK BROADENING OF Co II SPECTRAL LINES

19. Sladjana Marčeta Mandić, Luka Č. Popović, Edi Bon, Predrag Jovanović: SPECTROSCOPICAL INVESTIGATION OF GRAVITATIONALLY LENSED QUASARS

20. D. Marchev, Diana Kjurkchieva, S. Ibryamov, T. Atanasova: PHOTOMETRIC AND SPECTRAL INVESTIGATION OF THE STAR TU UMI

21. Anatolij A. Mihajlov, Vladimir A. Srećković, Nenad M. Sakan, Milan S. Dimitrijević: INVERSE BREMSSTRAHLUNG IN DWARF ATMOSPHERES: THE ABSORPTION COEFFICIENTS AND GAUNT FACTORS

22. Y. Nikolov et al.: POLARIMETRIC INVESTIGATIONS WITH TWO CHANNEL FOCAL REDUCER FoReRo2

23. Marko Pavlović, Dejan Urošević, Bojan Arbutina, Aleksandra Dobardžić, Milica Vučetić, Vladimir Zeković, Dušan Onić: RADIO EVOLUTION OF SUPERNOVA REMNANTS INCLUDING NON-LINEAR PARTICLE ACCELERATION

24. Rade Pavlović, Zorica Cvetković, Svetlana Boeva, Goran Damljanović: DETERMINATION OF ORBITS OF VISUAL BINARY AND LINEAR ELEMENTS OF DOUBLE STARS

25. N. I. Perov, A. A. Erokhin: WHERE DOES DUST MATTER EXIST IN THE SYSTEM OF PLUTO–CHARON?

26. Luka Č. Popović, Saša Simić, Nemanja Rakić: DETECTING SUBPARSEC SUPER-MASSIVE BINARY - LONG TERM MONITORING PERSPECTIVE

27. Miika Pursiainen, Predrag Jovanović, Luka Č. Popović: SHAPE OF IRON Kα LINE AND ACCRETION DISK PARAMETERS IN TYPE 1 AGN

28. Kristina Racković Babić, Dušan Marčeta, Stevo Šegan:THE MAYAN CALENDAR – FACTS AND DOUBTS

29. Dušan Raičević, Jovan Bajčetić, Vladimir M. Čadež, Aleksandra Nina: PROCEDURE FOR DETECTION OF CHARACTERISTIC RADIO SIGNAL VARIATIONS INDUCED BY SOLAR X-RAY FLARES

30. A. I. Shapovalova, L. Č. Popović, D. Ilić, V. Chavushyan, A. N. Burenkov, W. Kollatschny, A. Kovačević: OPTICAL SPECTRAL VARIABILITY OF QUASAR E1821+643

31. Saša Simić, Edi Bon: CONTRIBUTION TO SED OF AGNS INDUCED BY POSSIBLE DENSITY PERTURBATIONS IN COMPLEX GEOMETRY OF BINARY SYSTEMS

32. Vladimir A. Srećković, Anatolij A. Mihajlov, Ljubinko M. Ignjatović, Milan S. Dimitrijević, Darko Jevremović, Veljko Vujčić: HF ELECTRIC PROPERTIES OF THE ASTROPHYSICAL PLASMAS

33. Vladimir A. Srećković, Desanka M. Šulić, Anatolij A. Mihajlov: DISTURBANCES IN THE D-REGION INDUCED BY LARGE SOLAR FLARES

34. F. Taris, G. Damljanović, A. Andrei, J. Souchay, O. Becker: OPTICAL MONITORING OF QSOS IN THE FRAMEWORK OF THE GAIA SPACE MISSION

35. Ekaterina N. Tikhomirova: ASTRONOMICAL EDUCATION IN THE CULTURAL AND EDUCATIONAL CENTRE NAMED AFTER V. V. TERESHKOVA

36. Miljana Todorović Drakul, Jovan Bajčetić, Vladimir M. Čadež, Luka Č. Popović, Dragan M. Blagojević, Aleksandra Nina: REAL TIME VARIATIONS OF IONOSPHERIC TEC (TOTAL ELECTRON CONTENT) DURING SOLAR X-RAY FLARES

37. Milcho Tsvetkov, Damyan Kalaglarsky, Katya Tsvetkova, Darko Jevremović, Vojislava Protić - Benišek, Vladimir Benišek, Milan S. Dimitrijević: INCORPORATION OF THE BELGRADE OBSERVATORY ZEISS ASTROGRAPH PLATE ARCHIVE IN THE WFPDB

38. Doroteya Vasileva, Diana Kjurkchieva: LIGHT CURVE SOLUTIONS OF THE ECCENTRIC *KEPLER* BINARIES KIC 4281895 AND KIC 5115178 WITH TIDALLY INDUCED HUMPS

39. O. Vince, S. Samurović: BELISSIMA AND AS VIDOJEVICA: PLANS FOR THE FUTURE

40. M. M. Vučetić, B. Arbutina, M. Z. Pavlović, A. Ćiprijanović, D. Urošević, N. Petrov, D. Onić, A. Trčka: OBSERVATIONS OF NGC185 GALAXY – STUDY OF SUPERNOVA REMNANT IN A DWARF ELLIPTICAL GALAXY

41. M. M. Vučetić, A. Trčka, B. Arbutina, A. Ćiprijanović, M. Z. Pavlović, D. Urošević, N. Petrov: OPTICAL OBSERVATIONS OF NGC1156 GALAXY IN NARROW BAND [S II] AND Hα FILTERS

42. Krasimira D. Yankova: EVOLUTION IN THE SYSTEM CORONA-DISK

LIST OF PARTICIPANTS

Viktor L. Afanasiev
vafan@sao.ru
Special Astrophysical Observatory of the Russian AS, Nizhnij Arkhyz, Karachaevo-Cherkesia 369167, Russia

Jovan Aleksić
jaleksic@aob.rs
University of Belgrade, Department for Astronomy, Faculty for Mathematics, Studentski Trg 16, 11000 Belgrade, Serbia

Bojan Arbutina
bojan.arbutina@gmail.com
University of Belgrade, Department for Astronomy, Faculty for Mathematics, Studentski Trg 16, 11000 Belgrade, Serbia

Teodora Atanasova
t.atanasova@shu.net
Shumenski Universitet "Episkop Konstantin Preslavski", Ul. Universitetska 115, Shumen 9700 Bulgaria

Rumen Bachev
bachevr@astro.bas.bg
Institute of astronomy and NAO Rozhen,Sofia 1784, Blvd. Tzarigradsko shaussee 72, Bulgaria

Jovan Bajčetić
bajce05@gmail.com
Ruže Šulman 37, 23000 Zrenjanin Military Academy, Pavla Jurišića Šturma 33, 11000 Belgrade, Serbia

Vladimir Benišek
vlaben@yahoo.com
Astronomical Observatory, Volgina 7, 11060 Belgrade, Serbia

Robert Beuc
beuc@ifs.hr
Institut za fiziku, Bijenička 46 41000 Zagreb, Croatia

Nikolaj Genadievich Bochkarev
boch@sai.msu.ru
Universitetskij Prospekt 13, 119992, Moscow, Russia
Eurasian Astronomical Society, Russia

Svetlana Boeva
sboeva@astro.bas.bg
Institute of astronomy and NAO Rozhen, Sofia 1784, Blvd. Tzarigradsko shaussee 72, Bulgaria

Edi Bon
ebon@aob.rs
Astronomical Observatory, Volgina 7, 11060 Belgrade, Serbia

Nataša Bon
nbon@aob.rs
Astronomical Observatory, Volgina 7, 11060 Belgrade, Serbia

Daniela Boneva
danvasan@space.bas.bg, dany_sr97@hotmail.com
Space Research and Technology Institute, Acad. Georgi Bonchev Str., bl. 1, Sofia 1113, Bulgaria

Borislav Stanishev Borisov
b.st.borisov@abv.bg
University of Shumen, Faculty of Natural Sciences,
115, Universitetska Str., 9712 Shumen, Bulgaria

Duško Borka
dusborka@vinca.rs
Atomic Physics Laboratory (040), Vinča Institute of Nuclear Sciences, University of Belgrade,
P.O. Box 522, 11001 Belgrade, Serbia

Vesna Borka Jovanović
vborka@vinca.rs
Atomic Physics Laboratory (040), Vinča Institute of Nuclear Sciences, University of Belgrade,
P.O. Box 522, 11001 Belgrade, Serbia

Ivan S. Bryukhanov
betelgeize_astro@mail.ru
220033, Minsk, Sudmalis str. 16-27, Belarus

Vladimir M. Čadež
vcadez@aob.rs
Astronomical Observatory,
Volgina 7, 11060 Belgrade, Serbia

Yavor Chapanov
yavor.chapanov@gmail.com
National Institute of Geophysics, Geodesy and Geography (NIGGG), Bulgarian Academy of Sciences (BAS), Akad. G. Bonchev Str. Bl.3, Sofia 1113, Bulgaria

Aleksandra Ćiprijanović
aleksandra@matf.bg.ac.rs
University of Belgrade, Department for Astronomy, Faculty for Mathematics, Studentski Trg 16, 11000 Belgrade, Serbia

Zorica Cvetković
zorica@aob.rs
Astronomical Observatory,
Volgina 7, 11060 Belgrade, Serbia

Miodrag Dačić
mdacic@aob.rs
Astronomical Observatory,
Volgina 7, 11060 Belgrade, Serbia

Goran Damljanović
gdamljanovic@aob.rs
Astronomical Observatory,
Volgina 7, 11060 Belgrade, Serbia

Momchil Dechev
mdechev@astro.bas.bg
Institute of Astronomy,
72, Tsarigradsko Chaussee Blvd., Sofia, Bulgaria

Ivan Dimitrijević
ivand@matf.bg.ac.rs
Faculty of Mathematics, Studentski Trg 16, 11000 Belgrade, Serbia

Milan S. Dimitrijević
mdimitrijevc@aob.rs
Astronomical Observatory,
Volgina 7, 11060 Belgrade, Serbia

Vladimir Djošović
vladimirdjosowic@gmail.com
Society of Astronomers of Serbia,
Volgina 7, 11000 Belgrade, Serbia

Ana Djulaković
ana_djulakovic@yahoo.com
Velika Drenova 37245, Trstenik,
University of Belgrade, Faculty of
Physics, Studentski trg 12,
11001 Belgrade, Serbia

Sava Donkov
savadd@tu-sofia.bg
Department of Applied Physics,
Technical University-Sofia,
8 Kliment Ohridski Blvd,
1000 Sofia, Bulgaria

Robert Džudžar
Robert.Dzudzar@student.uibk.ac.at
Radnička 7, 25233 Ruski Krstur,
Serbia

Gantcho Gantchev
pgantcho@yahoo.fr
(Bul. James Bourchier 5, Sofia 1164,
Bulgaria), Department of Astronomy
Faculty of Physicsy,
Sofia University, Bulgaria

Sunay Ibryamov
sibryamov@shu.bg
University of Shumen, Department
of Theoretical and Applied Physics,
115, Universitetska Str.,
9712 Shumen, Bulgaria

Dragana Ilić
dilic@matf.bg.ac.rs
Department for Astronomy, Faculty
for Mathematics, Studentski Trg 16,
11000 Belgrade, Serbia

Natalija Janc
natalijanc@earthlink.net
415 Old Trail Road, Baltimore,
MD 21212, USA

Miljana D. Jovanović
miljana@aob.rs
Astronomical Observatory,
Volgina 7, 11060 Belgrade, Serbia

Predrag Jovanović
pjovanovic@aob.rs
Astronomical Observatory,
Volgina 7, 11060 Belgrade, Serbia

Damyan Kalaglarski
damyan.kalaglarsky@gmail.com
Institute of mathematics and
informatics, Bulgarian Academy of
Sciences, Sofia, Bulgaria

Evgeniya Alekseevna Karitskaya
karitsk@sai.msu.ru
Universitetskij Prospekt 13, 119992,
Moscow, Russia
Eurasian Astronomical Society,
Russia

Nikolay Kirov
nkirov@nbu.bg
New Bulgarian University,
Montevideo 21, Sofia, Bulgaria

Diana Petrova Kjurkchieva
d.kyurkchieva@shu-bg.net
Shumenski Universitet "Episkop
Konstantin Preslavski",
Ul. Universitetska 115, Shumen
9700 Bulgaria

Kostadinka Koleva
koleva@astro.bas.bg
Institute of Astronomy,
72, Tsarigradsko Chaussee Blvd.,
Sofia, Bulgaria

Ognyan Kounchev
kounchev@gmx.de
Institute of mathematics and informatics, Bulgarian Academy of Sciences, Sofia, Bulgaria

Andjelka Kovačević
andjelka@matf.bg.ac.rs
Department for Astronomy, Faculty for Mathematics, Studentski Trg 16, 11000 Belgrade, Serbia

Jelena Kovačević-Dojčinović
jkovacevic@aob.rs
Astronomical Observatory,
Volgina 7, 11060 Belgrade, Serbia

Lasko Laskov
lasko.laskov@gmail.com
Department of Informatics,
New Bulgarian University,
Montevideo 21, Sofia, Bulgaria

Georgi Yordanov Latev
glatev@astro.bas.bg
Institute of Astronomy with NAO Rozhen, Bulgarian Academy of Sciences, 72 Tsarigradsko Chausse Blvd., 1784 Sofia, Bulgaria

Igor F. Malov
malov41@mail.ru
142290, Pushchino, Moscow region, Pushchino Radio Astronomical Observatory, Russia

Sladjana Marčeta Mandić
sladjana.mandic@gmail.com
Astronomical Observatory,
Volgina 7, 11060 Belgrade, Serbia

Dragomir Marchev
d.marchev@shu-bg.net
Shumenski Universitet "Episkop Konstantin Preslavski",
Ul. Universitetska 115, Shumen 9700 Bulgaria

Gabrijela Marković
gbesic@yahoo.com
Faculty for Mathematics,
Studentski Trg 16, 11000 Belgrade, Serbia

Mihailo Martinović
Mihailo.Martinovic@obspm.fr
LESIA, batiment 16, 5, place Jules Janssen, 92190, Meudon, France

Žarko Mijajlović
zarkom@matf.bg.ac.rs
State Univ. of Novi Pazar, Novi Pazar, Serbia

Evgeny Mikhailov
ea.mikhajlov@physics.msu.ru
Moscow State University
119991, Leninskie gory, bldg.1, str.2, MSU, Faculty of Physics, Russia

Tatjana Milovanov
t_milovanov@yahoo.com
Astronomical Observatory,
Volgina 7, 11060 Belgrade, Serbia

Rosa Victoria Munoz Dimitrova
rdimitrova@astro.bas.bg
Institute of astronomy and NAO Rozhen, Sofia 1784,
Blvd. Tzarigradsko shaussee 72, Bulgaria

Petko Nedialkov
japet@phys.uni-sofia.bg
Department of Astronomy, Faculty of Physics, University of Sofia
5 J. Bourchier blvd, 1164 Sofia, Bulgaria

Grigor Nikolov
gnikolov@astro.bas.bg
Institute of astronomy and NAO Rozhen, Sofia 1784,
Blvd. Tzarigradsko shaussee 72, Bulgaria

Yanko Nikolov
ynikolov@astro.bas.bg
Institute of astronomy and NAO Rozhen, Sofia 1784, Blvd. Tzarigradsko Chausse 72, Bulgaria

Aleksandra Nina
sandrast@ipb.ac.rs
Institute of Physics, Pregrevica 118, 11080 Belgrade, Serbia

Vladimir N. Obridko
obridko@mail.ru
Euro-Asian Astronomical Society, Institute of Astronomy (Russian Acad. Sci.), and Sternberg Astronomical Institute (Moscow University), Russia

Vukan R. Ogrizović,
vukan@grf.bg.ac.rs
Department of Geodesy and Geoinformatics, Faculty of Civil Engineering, Belgrade, Serbia

Marko Pavlović
marko@matf.bg.ac.rs
University of Belgrade, Department for Astronomy, Faculty for Mathematics, Studentski Trg 16, 11000 Belgrade, Serbia

Nadežda Pejović
nada@matf.bg.ac.rs
Department for Astronomy, Faculty for Mathematics, Studentski Trg 16, 11000 Belgrade, Serbia

Nikolay Ivanovich Perov
perov@yarplaneta.ru
State Autonomous Organization of Culture "Centre named after V.V. Tereshkova", ul. Chaikovskogo, 3, Yaroslavl-150000. Russia

Nikola Ivanov Petrov
nip.sob@gmail.com
Institute of astronomy and NAO Rozhen, Sofia 1784,
Blvd. Tzarigradsko shaussee 72, Bulgaria

Velimir Angelov Popov
info@irida-observatory.org,
velimir.popov@elateobservatory.com
IRIDA Observatory, 57 Cherkovna str., Sofia, Bulgaria

Luka Č. Popović
lpopovic@aob.rs
Astronomical Observatory,
Volgina 7, 11060 Belgrade, Serbia

Vojislava Protić-Benišek
vprotic@aob.rs
Astronomical Observatory,
Volgina 7, 11060 Belgrade, Serbia

Kristina Racković Babić
rackovick@yahoo.com
Department for Astronomy, Faculty for Mathematics, Studentski Trg 16, 11000 Belgrade, Serbia

Dušan Raičević
ricek9@gmail.com
Military Academy, Pavla Jurišića Šturma 33, 11000 Belgrade, Serbia

Rajcho Roussev Rajsun
rajsun@abv.bg
Sofia, Bulgaria

Nemanja Rakić
rakinemanja@gmail.com
Faculty of Science, University of Banjaluka, Mladena Stojanovića 2, 78000 Banjaluka, Republic of Srpska, Bosnia and Herzegovina

Cyril Ron
ron@asu.cas.cz
Astronomical Institute CAS, Bocni II, 141 00 Praha 4, Czech Republic

Nikolai N. Samus
samus@sai.msu.ru
Institute of Astronomy (Russian Academy of Sciences), 48, Pyatnitskaya Str., Moscow 109017, Ryssia
Sternberg Astronomical Institute, Moscow, Russia and Euro-Asian Astronomical Society

Djordje Savić
djsavic@aob.rs
Astronomical Observatory, Volgina 7, 11060 Belgrade, Serbia

Zdravka Šejretova
japet@phys.uni.sofia.bg
Sofia, Bulgaria

Alla Shapovalova
ashap@sao.ru
Special Astrophysical Observatory of the Russian AS, Nizhnij Arkhyz, Karachaevo-Cherkesia 369167, Russia

Ivar Shmeld
ivarss@venta.lv
Ventspils University College, VIRAC Inzenieru 101 Ventspils LV-3600 Latvia

Saša Simić
ssimic@kg.ac.rs
Faculty of Science, Department of Physics, University of Kragujevac, Serbia

Zoran Simić
zsimic@aob.rs
Astronomical Observatory, Volgina 7, 11060 Belgrade, Serbia

Marijana Smailagić
marijana@aob.rs
Astronomical Observatory, Volgina 7, 11060 Belgrade, Serbia

Vladimir Srećković
vlada@ipb.ac.rs
University of Belgrade, Institute of Physics, P. O. Box 57, 11001, Serbia

Orlin Stanchev
o_stanchev@phys.uni-sofia.bg
Faculty of Physics,Department of Astronomy, University of Sofia 5 James Bourchier Blvd., Sofia 1164, Bulgaria

Milan Stojanović
mstojanovic@aob.rs
Astronomical Observatory,
Volgina 7, 11060 Belgrade, Serbia

François Taris
francois.taris@obspm.fr
Département SYstèmes de Référence
Temps-Espace, Observatoire de
Paris, 61, Av. de l'Observatoire
F-75014 Paris, France

Ekaterina N. Tikhomirova
en_tihomirova@mail.ru
State Autonomous Culture Institution
of Yaroslavl region «Cultural and
educational centre named after
V.V.Tereshkova»,
Tchcaikovskogo st., 3, Yaroslavl,
Russia, 150000

Miljana Todorović Drakul
mtodorovic@grf.bg.ac.rs
Department of geodesy and
geoinformatics, Faculty of Civil
Engineering, University of Belgrade
Boulevard of King Aleksandar 73,
11000 Belgrade, Serbia

Milcho Tsvetkov
milcho.tsvetkov@gmail.com
Institute of Mathematics and
Informatics, Bulgarian Academy of
Sciences, Acad. Georgi Bonchev
Str., Block 8, 1113 Sofia, Bulgaria

Tsvetan Angelov Tsvetkov
tstsvetkov@astro.bas.bg
Institute of Astronomy with
National Astronomical Observatory,
72, Tsarigradsko Chaussee Blvd.
1784, Sofia, Bulgaria

Katya Tsvetkova
katya@skyarchive.org
Institute of Mathematics and
Informatics, Bulgarian Academy of
Sciences, Acad. Georgi Bonchev
Str., Block 8, 1113 Sofia, Bulgaria

Dejan Urošević
dejanu@matf.bg.ac.rs
University of Belgrade, Department
for Astronomy, Faculty for
Mathematics, Studentski Trg 16,
11000 Belgrade, Serbia

Doroteya Vasileva
d.vasileva@shu.bg
Shumenski Universitet "Episkop
Konstantin Preslavski",
Ul. Universitetska 115,
Shumen 9700 Bulgaria

Todor Velchev
eirene@phys.uni-sofia.bg
Faculty of Physics, Department of
Astronomy, University of Sofia, 5
James Bourchier Blvd., 1164 Sofia,
Bulgaria

Oliver Vince
ovince@aob.rs
Astronomical Observatory,
Volgina 7, 11060 Belgrade, Serbia

Jan Vondrák
vondrak@ig.cas.cz
Astronomical Institute, Czech
Academy of Sciences, Boční II,
14100 Prague, Czech Republic

Milica Vučetić
mandjelic@matf.bg.ac.rs
Department for Astronomy, Faculty of Mathematics, Studentski Trg 16, 11000 Belgrade, Serbia

Krasimira D. Yankova
f7@space.bas.bg
Space Research and Technology Institute - BAS, Acad G. Bonchev Str., Bl. 1, 1113 Sofia, Bulgaria

Werner W. Zeilinger
werner.zeilinger@univie.ac.at
Department of Astrophysics at the University of Vienna, Tuerkenschanzstrasse 17, 1180 Wien, Austria

Igor Zinchenko
zin@appl.sci-nnov.ru
Institute Of Applied Physics RAS, Ulyanov st. 46, Nizhny Novgorod, Russia

AUTHORS' INDEX